The Sky at Einstein's Feet

William C Keel

The Sky at Einstein's Feet

 Springer

Published in association with
Praxis Publishing
Chichester, UK

Professor William C Keel
Department of Physics and Astronomy
University of Alabama
Tuscaloosa
Alabama
USA

SPRINGER–PRAXIS BOOKS IN POPULAR ASTRONOMY
SUBJECT *ADVISORY EDITOR*: John Mason B.Sc., M.Sc., Ph.D.

ISBN 10: 0-387-26130-3 Springer Berlin Heidelberg New York

Springer is a part of Springer Science + Business Media (*springeronline.com*)

Library of Congress Control Number: 2005932442

Cover design: Jim Wilkie
Copy editing and graphics processing: R. A. Marriott

Printed in Germany on acid-free paper

For all those who have been unjustly forgotten after a century

Preface

The year 2005 will see many appreciations of Einstein's legacy, but is is sheer happenstance that this book came along now. It was inspired by the Washington status of Einstein, which stresses his connections to the cosmos. That line of thought led to this volume of explanation, which I hope will share with a wider audience the deep impact of relativity on astronomy. Over and over, we have found things happening in the Universe which required the insights of relativity to explain. Mnahy of these deserve to be lifted from the research literature and graduate lectures to a wider public.

Many people have generously helped me with images, data, and personal recollections. I would like to acknowledge with thanks the narratives provided by Ray Weymann, Bob Carswell, Pamela Walsh, Barbara Schaefer, Jackie Hewett, Alan Stockton, Bob Goodrich, Derek Wills, and Roger Lynds on gravitational lensing, and Don Schneider, Myron Smith, Jack Sulentic for recollections of Peter Young and Richard Wilcox. Jay Holberg allowed me to see his unpublished manusctript on white dwarfs, and his detective work on how Adams got the wrong answer by doing all the right things. Don Osterbrock, Richard Dreiser, and Kyle Cudworth helped me try to straighten out early responses to Einstein's preduction of gravitational lensing. Jerry Fishman, Bruce Margon, Phil Hardee, and Felix Mirabel provided useful (and sometimes quotable) comments. Rem Stone and Tony Misch at Lick Observatory kindly checked details of the plate archives to work out who observed what a century ago. Howard Bond, Mel Ulmer, Riccardo Giacconi, Laurent Bouchet and William Griffith provided pictures. Charles Alcock and Pavlos Protopapas provided the pointers I needed to present some of the MACHO results, and Jessica Moy of ther NOAO library tracked down information from their plate logs. I caused Ann Wehrle and Glenn Piner more grief than I intended in producing a pleasing rendition of superluminal motion in a radio source. Steve Mencinski provided helpful comments from reading an early draft. I also acknowledge the tremendous service provided by the agancies which aintain public data archives which are indispensible to so many of us – particularly NASA, the European Space Agency, and the Center for Astrophysics. Even if we're not giants, we can reach very high if we all share our smaller steps.

I am grateful to Terri for proofreading and so much more.

Tuscaloosa, Alabama,
William C. Keel
May 2005

Contents

List of Figures

Colour Tables

1 The sky at Einstein's feet

Among the many monuments in the city of Washington, there is one only a block from the National Mall which offers a more human scale than the better-known commemorations of the sacrifices of war and the drive and inspiration of presidents, while giving the imagination wings to span the Universe. In front of the National Academy of Sciences, amid a grove of elm and holly trees that has made it a pleasant diversion for countless groups of schoolchildren during sultry Washington summers, is a statue of Albert Einstein. Sculptor Robert Berks has depicted him seated in thought, holding a tablet as if to echo traditional depictions of Moses bearing the tablets of the Law. Einstein's single tablet also depicts law – three equations characteristic of Einsteins's major contributions to physics. Two of them date from his "miracle year" of a century ago. That year, 1905, saw his publication first of his explanation of the emission of electric current from certain metals (the photoelectric effect) in terms of the particle-like behavior of light – work which was the primary driver for his Nobel Prize in 1921 – followed quickly by the first revolution of relativity. This paper, "Zur Elektrodynamik bewegter Körper" or "On the electrodynamics of moving bodies", hid within a dry and pedestrian title a content which was explosive in both literal and metaphorical senses for physics, astrophysics, philosophy, and even the world balance of power. The tablet (Fig. 1.1) recognizes this with the most familiar equation in all of physics: $E = mc^2$. The middle line of the tablet shows the relation between the energy of incoming radiation and electrons liberated by the photoelectric effect. The uppermost equation comes from the second of Einstein's revolutions, the formulation of general relativity. This work appeared only in 1916, after a digression as Einstein mastered the new mathematical tools it required. The bottom equation on the tablet relates the curvature of space to the density of mass and energy – a discovery which superseded the physics of Newton, after two centuries, with an understanding which is at once more accurate and vastly stranger in concept.

But this stone Einstein (in German, *ein Stein Einstein*) pays no regard to his equations. Instead, he gazes downward in thought, toward a granite representation of the sky as it stood overhead when the monument was dedicated on April 22, 1979. This connection of earth and sky has an honorable tradition – there is a similar depiction atop Hoover Dam, itself in a line stretching back to ancient times. But something is different here, uniquely reflecting Einstein's vision and achievements. This sky in metal and stone depicts not only the stars and planets of human vision, but adds quasars, pulsars, and the other citizens

of the Universe which we have begun to understand only with the tools Einstein
left us. This is a rich legacy – one which has continued to deepen with new
discoveries that Einstein himself did not live to see. If anything, his legacy to
astronomers is deeper than the legacy he left for physicists.

Fig. 1.1. Science-minded traveler Nathan Keel with Einstein's stone tablet. The equations above Einstein's signature embody three of his greatest contributions to physics (and to astronomy). From the top, they describe the curvature of space induced by mass–energy, from general relativity; the relationship among the energy of a photon, the ionization energy of atoms, and the energy of an escaping electron, from his 1905 description of the photoelectric effect; and finally, the familiar mass–energy equivalence formula from special relativity.

To be sure, Einstein has become an icon without peer for physicists. The twin theories of special and general relativity set the gold standard for scientific ideas – mathematically elegant, completely different from what came before, and predicting new phenomena in the broadest imaginable range of situations. To this day, doing a calculation relativistically means not only including all the effects predicted from relativity and not by "classical" physics, but has taken on the informally added connotation of doing everything right so the results will hold in all conceivable circumstances. Einstein had a particular genius (if one may be singled out) for the thought experiment (*Gedankenexperiment*)– an imaginary situation which self-evidently embodies a physical principle and shows how certain theories either lead to paradoxes or explain an unexpected result clearly. Einstein's persona also gave us three generations worth of stereotypes of the theoretical scientist – a gentle and other-worldly demeanor, a genial shabbiness, an indifference to social custom. When the writers of the script for *The Day the Earth Stood Still* wanted to depict a brilliant scientist with world-wide renown, they went for a character with thin and wild hair, a blackboard covered with equations, and a distracted manner. When Frank Oz wanted to give the Jedi master Yoda an air of ageless wisdom and compassion, he patterned the puppet's eyes directly after a photograph of Einstein. Chasing after the image applies even to attire – generations of scientists have patterned their dress after the rumpled sweaters and sandals Einstein preferred, to the despair of family and image-conscious administrators alike. The image has permeated our society – Einstein appeared on the cover of *Time* magazine no less than five times, once as Person of the Century.[1] His visage has appeared on postage stamps from many countries (Fig. 1.2), as well as posters, T-shirts, and even an action figure available from various Internet dealers. Einstein's persona is so evocative that it shows up in the most unexpected venues, invariably attracting attention (Fig. 1.3).

However long his shadow on physics, it stands so long and so broad for astronomy and astrophysics that we scarcely even notice that we are standing in it. I realized after seeing the statue in Washington how uniquely appropriate it was, this depiction of the sky at Einstein's feet. Astronomers have lived for decades in the Universe that Einstein unveiled, and have seen new discoveries over and over illuminated by the insights of relativity. It is in the cosmic arena that we find matter moving close to the speed of light, masses so huge that they visibly distort space and bend starlight, illusions caused by the speed of light scarcely outrunning the material that emitted it. This is where we find our existence dependent, in the cosmic past and minute by minute, on the physical expression of $E = mc^2$ as it tells us why the stars shine. And it is here that we confront the ultimate mysteries of black holes and cosmic origins.

---

[1] These were February 18, 1929; April 4 1938; July 1, 1946, as "cosmoclast" with a mushroom cloud in the background; Feb 19, 1979, commemorating the centennial of his birth, and as Person of the Century, December 31, 1999. In addition, Elsa Einstein appeared on the *Time* cover for December 22, 1930.

Fig. 1.2. The ubiquity of Einstein's image is illustrated by his appearance on this 1979 postage stamp from the United States, one of many such appearances worldwide. (Copyright US Postal Service.)

The book explores the Universe that we have uncovered through the eyes Einstein gave us. Some of its aspects may be familiar; others have been seen mostly in the pages of technical journals and the online world of preprint archives. Pieces have been revealed with successively larger telescopes on Earth, the great dishes of radio astronomers, and a fleet of spaceborne observatories of ever more exquisite precision and sensitivity. But most of all, they stand revealed through

Fig. 1.3. Einstein's reputation forms the backdrop for this piece by folk artist Butch Anthony, from his 'Museum of Wonders" series. (Reproduced by permission of the artist.)

the understandings of space, time, matter, and energy that are parts of Einstein's legacy.

To set the stage, we will follow a timeline of some of the key discoveries of the last century. At the opening of the twentieth century, astronomy was slowly undergoing the first of a series of revolutions – the rise of astrophysics. Astrophysics, at its outset, was the application of laboratory knowledge about the spectrum to interpretation of the light of stars and nebulae, although its boundaries and relation to astronomy have blurred enormously since then. This confluence of the laboratory and the observatory set the stage for learning what the stars are made of, how they shine, and the revelations of cosmology to come. But advances in physics were yet to give our predecessors a century ago the keys to the mysteries of the stars. They had as yet no coherent description of why stars worked, much less how the zoo of stars in the night sky might be related to each other, or could be born, exist, and die. The form of the Milky Way holding these stars was only dimly sketched, and the existence of other galaxies beyond it remained a matter for the crudest speculation. Newton's laws of motion and universal gravitation, formulated in the seventeenth century, still reigned in the heavens, producing the exquisitely accurate descriptions of the planetary motions that had long ago made celestial mechanics the first of the truly exact sciences. Chronologically, Einstein's publication of special relativity in 1905 was one of the first hints that a veil was being pierced, but initially he had trouble in interesting astronomers in pursuing applications of this abstruse theory. Once Einstein had worked out general relativity, though, astronomers could not hide – with its explanation of subtle changes in the orbit of Mercury, relativity had taken on the hallowed ground of celestial mechanics, and the gravitational deflection of starlight predicted to be seen during eclipses was a new and utterly unexpected astronomical phenomenon, attracting the attention of some of the leading astronomers of the day.

By 1925, we were learning what the stars are made of, to the considerable surprise of physicists and astronomers alike. The key step was made by Cecilia Payne (later Payne-Gaposchkin), who first learned the newly unfolding quantum physics in Cambridge, England, and went on to apply these lessons to the spectra of stars, at Harvard, in another Cambridge. Taking into account which spectral lines should appear at various temperatures, she found that most of the bewildering variety of spectra seen for stars reflected a change in temperature alone, so that most stars were made of roughly the same mix of elements. However, her calculations showed that this mix was throughly dominated by the lightest elements, with hydrogen and helium accounting for about 99% of the gas in the Sun and similar stars. This result, which was quite correct, was so deeply at odds with contemporary ideas about the stars that her dissertation advisor was deeply skeptical, and in 1925 Payne-Gaposchkin herself would still write that the "stellar abundance deduced for these elements is improbably high, and is almost certainly not real". [2] But they were real, and it was this knowledge that would lead a decade later to our understanding of what powers the stars.

Another path from the laboratory was leading us to the answer to this puzzle. The mysteries of radioactivity had riveted many of the world's leading scientists for decades since Becquerel had found his photographic plates fogged by radium in 1896. The seeds of our understanding of fusion processes in stars may be said to date to 1920, in a series of measurements by F.W. Aston designed to compare the masses of atoms of various elements. Aston remarked only in passing on the fact that four hydrogen atoms weigh 0.7% more than a single helium atom. However, Arthur Stanley Eddington, perhaps as familiar with the consequences of relativity as anyone beyond Einstein himself, saw quickly that this mass deficit might be translated into energy within stars. By the eve of the Second World War, Hans Bethe in the US and Carl von Weizsäcker in Germany had sketched

---

[2] *Proceedings of the National Academy of Sciences*, 11, 3, p. 197, available electronically from the *www.pnas.org* site.

out the most important pathways for nuclear fusion likely to occur in the interiors of stars. With the outbreak of war, research on nuclear processes received enormous impetus, with results both great and terrible.

At the same time, a handful of astronomers – above all, Edwin Hubble – were taking advantage of the new giant telescopes in California to reveal the Universe beyond our galaxy – and indeed that it made sense to have a plural form of "galaxies". Hubble's first big splash was the demonstration that some of the brightest "white nebulae", including the large spiral in Andromeda, contain variable stars of a well-known kind, whose faintness meant that they must be at distances of hundreds of thousands of light-years (and more), and thus located in systems genuinely comparable to the Milky Way. This was followed within a few years by Hubble's measurements of spectral shifts (looking like the familiar Doppler shift) for a few dozen galaxies, and (in a somewhat lucky guess, in view of the sparse data he had in hand) surmised that more distant galaxies exhibit uniformly greater spectral shifts. [3] This led, in short order, to the picture of the expanding Universe – one which could be incorporated into a cosmological picture firmly rooted in general relativity. Einstein saw this at odds with his previous assumption of a static Universe, and retracted his inclusion of a stabilizing term in one of his equations. Seventy years later, we find that he may have been right the first time.

Among the technological spinoffs by war's end had been enormous improvements in electronics, particularly transmission and receiving equipment related to radar. The availability of this equipment, and the experience many physicists had gained in war's crucible, brought radio astronomy out of the backyard status it had reached in the 1930s [4] and allowed the exploration of a completely new window on the Universe. This window showed us quasars, pulsars, and radio galaxies, whose main radiation processes operate only in domains of energy and velocity so great that relativity plays a dominant role. Radio telescopes of increasing size and sophistication revealed jets of material shot away from tiny objects at nearly the speed of light, complete with the optical illusions induced by such motions, long described in relativity books but hitherto remaining only abstractions. From the 1950s into the 1970s, radio astronomers were in the vanguard of showing us the "violent Universe", a far cry from the serenity of the starry sky we had always known – or thought we knew.

Using surplus vacuum tubes after World War II, astronomers could make widespread application of another of Einstein's brainchildren – the photoelectric effect, now understood in detail – to add precision to their measurements. Tubes initially designed to generate electronic "noise" for radio jamming could be used in the other direction, to measure the amount of incoming light from a star or

---

[3] Hubble had not actually discovered the redshifts of galaxies – that honor probably belongs to V.M. Slipher, working with the 24-inch refracting telescope at Lowell Observatory – but he did substantially extend the amount of data available, and systematized it to a point allowing this breathtaking generalization

[4] Quite literally – Grote Reber was operating a 36-foot dish antenna at his Illinois home during the 1930s, being by several decades the first homeowner to have neighbors wondering what that thing was.

galaxy with unprecedented accuracy. This practice of photoelectric photometry was the pinnacle of technical sophistication for a generation of astronomers,
surpassed in the 1980s by the more advanced technology of charge-coupled devices (CCDs), which now applied the photoelectric effect to millions of pixels at
a time, in a precise geometric array. Precise photometry provided data on the
sizes and temperatures of stars by the thousands, becoming the basic data for
understanding the life cycles of stars.

The Universe became an even more interesting place with the advent of X-
ray astronomy, a child of the space age. Starting in 1962, a series of rocket and
satellite payloads revealed flickering emission from intensely hot matter near
neutron stars and black holes, becoming the technique par excellence to find
black holes. X-rays also emerge from the powerful beacons of quasars, part of
the evidence that they mark the most massive and influential black holes.

By 1980, we had the technology and luck to discover gravitational lenses, the
logical extension of the deflection of starlight through a gravitational field which
caused such a stir in 1919. Twenty-five years later, this is a vital research area
in its own right, furnishing us with ways to map mass, seek planets, measure the
images of distant stars, and extend the range of our telescopes.

Astronomers found themselves exploring Wonderland in the 1990s, the era
of the orbiting *Hubble*, *Compton*, and *Chandra* observatories, and of the giant 8-
and 10-meter telescopes sprouting in Hawaii and Chile. The motions of stars in
the hearts of galaxies betrayed enormous black holes not just in quasars, but in
virtually every bright galaxy (including our own). Enigmatic flashes of gamma
rays have been tracked to vast explosions in the depths of space and time, flinging
material outwards at nearly the speed of light and beaming energetic radiation
across the Universe. Long-term programs designed to measure important parameters of relativistic cosmology returned the unexpected (and quite shocking)
result that the cosmic expansion is accelerating, perhaps a vindication after all
these decades of Einstein's notion of a "cosmological constant", a theoretical
notion that he had rejected for lack of empirical evidence.

Relativity has pervaded astronomy so deeply that this tour will be a bit
like a review of the greatest hits of twentieth-century astronomy. I have had to
pick some way to organize this journey, and picked one theme for grouping the
various astronomical topics – which of the facets of relativity they express. This
entails some jumping back and forth – but be assured that astronomers jumped
around a lot more in the quest for understanding. Some of these historical jumps
will be evident as I trace the history of discoveries, often not just dependent
on personalities but embodying utterly unlikely people and circumstances. The
practice of science is many things, but bloodless and coldly rational it is not. Nor,
being an activity practiced by humans, can it be. This book is a personal journey
through Einstein's cosmic legacy. In some fields, I am at home; in others, we are
fellow tourists. The particular topics and historical threads are also a personal
selection; some especially interesting roads are followed, others not taken for
reasons having nothing to do with the importance of the work done there. I
must apologize at the outset to all those who, because of space limitations or

my own narrative direction, are not named in the text for their contributions in the laborious construction of the roads we explore.

One textbook on relativity in astrophysics cautions readers in its first pages that "The equations that give these hard-to-believe and challenging results are easy to write down but equally easy to use incorrectly". [5] It is challenging, and somewhat risky, to describe the relativistic Universe in words rather than symbols. It is very easy to stray from the path and get thoughts tangled in this way, rather than using the abstract mathematical language which is what we seem to share most fundamentally with the physical Universe. In important ways, this book is a work in translation – but translations are the way we begin to appreciate new languages.

---

[5] W.C.S. Williams, *Introducing Special Relativity*, Taylor and Francis (London, 2002).

2 Bookkeeping at the speed of light

The speed of light is the speed of light. Period, full stop. Any time we are dealing with it in a vacuum, when it is not slowed down by interactions with matter. That is the basic message of Einstein's special relativity. The speed of light, and of all its kin throughout the electromagnetic spectrum from gamma rays to radio waves, is what it is, and is unchanged by any motion of either the source or receiver of the light. If something has to give in seeing how this can be so even with objects moving at unearthly speeds, it will be our classical, intuitive notions of how space and time behave. The speed of light is more than a number describing how fast a particular phenomenon goes – it is a value woven into the deepest fabric of our Universe.

2.1 Looking out is looking back

Some astronomical effects of the speed of light are no more complicated than doing the bookkeeping as to what we expect to see when; less obvious effects of relativity start to sneak in as we look farther and farther across the Universe, or deal with motions that are not tiny compared to the speed of light. This speed is enormous by all our everyday standards – nearly 300,000 kilometers/hour (186,282 miles/second), universally denoted by c.

That we see everything in the sky as it was some time ago is a fundamental part of the way we perceive the Universe, and indeed, since our senses and instruments are mediated by radiation, it must be so. This is one of the cues used by a GPS system to determine position, starting from the continuously updated locations of the satellites emitting the navigation signals. It shows up in long-distance telephone calls whenever they have been routed through satellite circuits rather than land lines, as the round trip to a satellite 35,800 km above the equator produces an extra 0.5 second of delay between sides of a conversation – just enough to feel awkward with someone we know well. The delays get longer and the news gets older as we look farther afield. Conversations with astronauts on the lunar surface included delays of about 2.7 seconds, forcing even the President of the United States to wait on the radio waves during the "most historic telephone call ever made". Engineers and mission scientists holding their breaths awaiting the fate of the twin Mars rovers in early 2004 ran low on oxygen for an extra 9 minutes even with Mars rather close, and the carrier waves from the European Space Agency's Huygens probe (which has just entered the atmosphere of Saturn's huge moon Titan as this is written) took nearly an hour and a half to reach us. It was the vast span of the Solar System which led Ole Rømer to conclude in 1676 that light does have a finite speed, rather than arriving instantly at its destination. He based this on noting that the moons of Jupiter appeared to enter and leave eclipse by the planet too early at some times of our year and too late at others, in a pattern which exactly matches changes in Jupiter's distance from Earth. Galileo had attempted such a measurement using lamps on hilltops, but it took what Isaac Asimov described as two hilltops half a billion miles apart to show the delay of light speed in an unmistakable way.

Our familiar Solar System is crowded and urban when compared to interstellar space. The light we see from the three stars of the α Centauri system is over four years old by the time we see it, and there are a few stars visible to the naked eye which we see as they were before the dawn of recorded history. We trace the decades-long orbits of stars around the black hole at the heart of the Milky Way as they happened 24,000 years ago, and watched the explosion of the star Sanduleak -69° 202 about 160,000 years after the fact. [1] But these time spans are trivial compared to the scale for change on really large scales. Even the gulf between here and the Andromeda galaxy can be spanned by light in only the time it takes either spiral to turn only a few degrees, so neither one would look much different across the interim (barring the 10,000 or so massive stars which had either been born or exploded in that time). It is in staring into the abyss of deepest space that we also gaze into deepest time. As we see galaxies billions of light-years away, we look back to times early in the history of the Universe itself, when the galaxies were so much younger that their complements of stars were different, their enhancements of heavy elements incomplete, and their structures had yet to settle down to today's symmetric spiral and elliptical forms. Distance and time are always mixed in our view, so that telescopes cannot help being one-way time machines, and giving us a gift in our study of the past which is denied to ordinary historians and palaeontologists. As galaxy-evolution specialist Alan Dressler puts it, by working hard enough we can see someone else's past – and in cosmic terms that is just as good as our own. Space and time are inextricably linked not just in the formalism of relativity, but in any astronomical observation. Although applied to a rather different world, Tolkien's mythology phrased the importance of both parts of our position precisely. Before the First Age of Middle–Earth, Eä, the World which Is, was created "in the Deeps of Time and amidst the innumerable stars".

---

[1] Astronomers have picked up the habit of implicitly quoting properties of distant objects simply referenced to whatever time the object's light originated in, rather than folding in the light–transit time, probably from the clumsiness and pedantry of constantly writing "a galaxy whose age is estimated at four billion years based on light which has been in transit for a computed three billion years, meaning that, in the impossible case that we could see it at the time equivalent to now in a reference frame at rest in the cosmic expansion, we would see it at an age of seven billion years". This habit has carried over into press releases, where it can sometimes lead to sublime confusion in understanding new results about galaxies in the early Universe.

2.2 Looking sideways to catch the light

As a graduate student, I compiled an atlas of stellar spectra, covering a wide range of temperature and luminosity. These were to be used to numerically match galaxy spectra as combinations of stars, so I could look for tiny deviations that represented weak emission from gas in their centers. One of the red giants I included was γ Draconis, which had all the features I was looking for. It was bright, so there was no problem observing during full Moon – as it was, I had to insert a gray filter in front of the spectrograph. It had been well observed and had a secure spectral type and color. And it was conveniently placed in the northern sky to be observable from Mt. Hamilton for much of the year, but not so far north that the telescope would bump into its yoke mounting. This star has proven popular with observers for such reasons, both for direct observation and for calibration. NASA's Astrophysics Data Service lists 371 published papers involving γ Draconis. But its place in the history and lore of astronomy was already secure in 1727. This star gave us the first direct evidence that the Earth moves.

To be sure, astronomers did not doubt the motion of the Earth, two centuries after Galileo. But the arguments had been indirect and the evidence circumstantial. Yes, the phases of Venus showed that at least one planet orbited the Sun; and the moons of Jupiter illustrated that other bodies could be the centers of motion. After Galileo, Italy's Jesuit astronomers found that the Sun's apparent size changes during the year in just the way predicted by Kepler, not Ptolemy. But the real prize would be proof – measurement of the shift in a star's apparent position as we observe it from different parts of the Earth's orbit each year. The quest to see this effect occupied some of the most careful and inventive astronomers for over a century. Among them was James Bradley, who used a long telescope pointing near the zenith (a so-called *zenith sector*, still on view at the National Maritime Museum in Greenwich) near London. This arrangement minimized errors due to atmospheric refraction, and allowed use of very precise scales over shorter angles than other designs. He chose γ Draconis as his quarry, because it was a bright star (hence, possibly close to us) which passed close to the zenith from his latitude. Bradley did find an annual shift in the star's location, as it traced out an elliptical path in celestial coordinates with a major axis of 40 arcseconds. As he soon realized on comparison with other stars, which showed a matching shift, he had found not the expected parallax, but something intellectually just as powerful in demonstrating the motion of the Earth – the aberration of starlight.

If we observe light from some distant object, reaching us at c, from a platform whose motion changes in direction, we will have to point our telescope slightly "upstream", just as an umbrella must be tilted forward to catch the rain if we run through it. The amount of this tilt is set by the ratio of the orbital speed of the Earth to c, and for any star the tilt will change depending on the time of year. For stars near the pole of the Earth's orbit, the apparent path is a circle, while for stars lying in the plane of the Earth's orbit, they follow a one-dimensional arc back and forth in our coordinates.

Sir John Herschel pointed out, in the mid-nineteenth century, that all stars exhibit the same aberration shifts, regardless of the motions revealed by delicate positional measurements. This led Herschel to propose, well before Einstein's work, that the velocity of light must be the same regardless of the relative motion of Earth and stars (although the point could not be made very strongly using the stars whose motions were known at that time). This point is made vividly by every long-exposure *Hubble Space Telescope* image, such as the Hubble Deep Field. These data include not only the aberration due to the Earth's orbit around the Sun, but the aberration associated with the telescope's orbit around the Earth. This makes the images sway back and forth by about 9 arcseconds during each 93-minute orbit. This amount changes very slightly from one edge of the telescope field of view to the other, as one edge looks more nearly "upstream", and sets a limit to the precision of tracking with two instruments at the same time. In the Deep Field, there are objects spanning the enormous redshift range from stars in our galaxy, with radial velocities of only tens of kilometers per second, to galaxies at redshifts[2] greater than $z = 5$, which are receding from us at nearly the speed of light.[3] All these show quite precisely the same amount of aberration; if they did not, each image would show different amounts of blur for various objects, and long-exposure stacks of many images taken over a period of time would be blurred beyond recognition.

By now, astronomers scarcely notice aberration except for measurements pushing the limits of precision. It is routinely removed in telescope pointing by computers, on the plausible theory that computers should compute so people can think. It shows up in radio interferometry in a slightly different guise – subtle changes in the arrival time of equivalent waves as the whole array of antennas has moved during the time it took for the waves to travel between them.

2.3 Echoes in Einstein's Universe

Much more spectacular ways to see the speed of light in action are furnished by light echoes. When a star explodes, or suddenly brightens a thousandfold for a less fatal reason, the flash can illuminate dust and gas in its immediate surroundings, as well as spread out along the whole line of sight in front of the star. We see this as an expanding bubble of illuminated material, which seems to expand far beyond the speed of light. Ordinary intuition tells us that the bubble, being the amount of dust lit up by an astronomical flashbulb, expands

---

[2] Astronomers denote the measured redshift of an object with the ratio z of wavelength shift to the emitted, or laboratory, wavelength. In this case, for $z = 5$, all observed spectral features are at wavelengths $1 + 5 = 6$ times as large as when they left the quasar. On cosmic scales, redshifts imply light-travel distance and lookback time to a source. Roughly, $z = 0.5$ corresponds to 5 billion light-years, $z = 1 - 8$ billion, $z = 2 - 10$ billion, and $z = 5$ reaches 12.5 billion years back.

[3] Or more precisely, the expansion of space between then and there, and here and now, is increasing the distance, without either one having to be "moving" in the usual sense of the word.

at the speed of light. But our intuition fails us here, as that is not what we are seeing! Instead, at each instant, we see the dust which is in the right place for the time since the flash to equal its travel time first to the dust, and then at an angle, on toward the Earth. This satisfies the geometric definition of an ellipsoid – the shape for which the sum of distances first from one focus, then to the other focus, is a fixed constant. At any time, we will see reflected light from whatever dust grains happen to lie on the surface of a very long ellipsoid stretching from behind the star to behind the Earth. Dust in front of the star is most effective in redirecting starlight, so that is what we usually see. Light echoes were first seen around star systems undergoing nova outbursts – starwide nuclear explosions that can happen when a companion star dumps too much fresh hydrogen onto the surface of a dense stellar remnant known as a white dwarf.[4] The math had all been worked out by French astronomer Paul Couderc (1899–1981), who was noted not only for his astronomical research but for his efforts in promoting public understanding of science (and particularly relativity). In a 1939 paper which still gets waves of citations, he explained the so-called light echoes which had been photographed around Nova Persei 1901. Each ring marks the intersection of a cloud of interstellar dust, scattering the light into our view, with an ellipsoidal surface defined by the time since we saw the illuminating outburst (Fig. 2.1). The ringlike light echoes can appear to expand much faster than the speed of light. This seems paradoxical after learning that special relativity tells us that nothing can exceed the speed of light. Light echoes do not, of course, actually violate any tenets of relativity since nothing material is really moving so fast – we see only geometrically defined locations, somewhat like a searchlight beam sweeping along a wall.

Many astronomers rediscovered light echoes while observing the aftermath of the nearest supernova to be seen since Galileo first turned his telescope skyward. In February of 1987, a naked-eye star blazed in one of the Milky Way's satellite galaxies. This galaxy, the Large Magellanic Cloud, lies about 160,000 light-years away, so far south in our sky that only observers south of the latitude of Mexico City can see it at all.

Supernova 1987A was seen to explode on February 23, 1987. The formal discovery was made by Ian Shelton, an observing assistant at the southern station of the University of Toronto at Cerro Las Campanas in Chile. He was investigating use of an otherwise obsolete lens system for wide-field pictures by taking a photographic plate of the Large Magellanic Cloud. On examining the plate in the darkroom, while it was still wet from processing, he noticed a bright star image near the center, and at first wondered why he had not selected that star as a reference for the guiding eyepiece. By this time, Oscar Duhalde had noticed the star visually, while taking a break from operating one of the larger telescopes at the site. Shelton notified the other two astronomers at the site, and the four went outside to gaze at the first supernova to be seen by unaided

---

[4] Confusing the issue, nova outbursts do shoot off rapidly expanding shells of gas – which become visible much later, travelling at only a tenth of a percent of the speed of light.

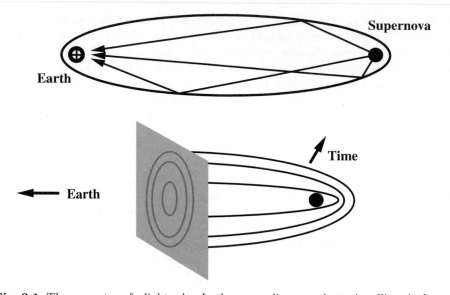

Fig. 2.1. The geometry of a light echo. In the upper diagram, the entire ellipse is shown. We will potentially see a light echo from all points along this ellipse (in fact a three-dimensional ellipsoid depicted in cross-section). As the arrows indicate, all these points share the same light-travel time to Earth. The lower diagram shows the expansion of the ring from a sheet of dust as time passes, in a view schematically magnified near the exploding light source. As time passes, the ellipsoidal location of echoes expands, and we see light where it intersects clouds containing dust.

human eyes since 1604. They then turned attention to communicating the event to the worldwide astronomical community, which was more than a simple matter of email even as recently as 1987. They sent a message to the International Astronomical Union's Central Bureau for Astronomical Telegrams at Harvard, which received a confirming report from New Zealand only a half hour later. In hindsight, Robert McNaught in Australia discovered that a series of routine photographs he had taken on the previous night showed the supernova's initial brightening before it reached naked-eye visibility, while inveterate variable-star observer Colin Henshaw, working in East Africa, noted and photographed the supernova as night reached Zimbabwe, but was in no position to send a report for days.

Happening in one of our satellite galaxies, a system which has been closely surveyed as a laboratory for the life cycles of stars, Supernova 1987A was the first supernova of which we knew something about the star before it blew up. The event marked the demise of a brilliant blue star which had been catalogued as Sanduleak -69° 202, on the edge of the great stellar factory of the Tarantula Nebula. This event was also observed by neutrino detectors intended to monitor the Sun, providing deep insight into how stars explode. But for the moment, our interest is in the way the titanic brightness of the explosion lit up material around the star.

The dying star shone about as bright as the North Star for several weeks, gradually fading to reveal its surroundings. These surroundings included a set of rapidly expanding rings which were quickly identified as its light echoes, offering a unique chance to map the dust in this nearby galaxy in three dimensions (Fig. 2.2). Each ring showed where the expanding ellipsoid from which the light reached us after the right time intersected a cloud or sheet of interstellar dust. [5]

A recent study, published by a team of astronomers spanning the distance from New York to Chile, has combined results of 16 years worth of light–echo observations to map the surroundings of Supernova 1987A. They found that the star's impact on its surroundings, well before its final explosion, extended light-years beyond the glowing "hourglass" nebula that appeared so striking when *Hubble* was first turned in its direction. The star had been blowing a powerful wind throughout most of its lifetime, in at least three different phases, leading to a huge peanut-shaped cloud of surrounding dust and gas stretching 28 light-years outward along each pole and 11 light-years near the equator. The nebula we see glowing by gaseous emission in the usual images has rings only 1–1.3 light-years from the star – most of its previous mass output stands revealed only briefly, as it reflects the blinding flash of the star's end.

It seems odd that we see dust so far from the supernova reflecting its light so brightly. The secret is that interstellar dust particles are tiny, almost all less than a quarter micron (that is, 0.00025 millimeter) in size. Such fine particles (like fine hair, for that matter) direct light most effectively forward, deflecting it only slightly (known as forward scattering). Therefore, most of the light from the supernova bounces off particles of interstellar dust at angles of only a few degrees to its original direction, making it easiest for us to see the dust that lies almost in front of it (as well as the natural dominance of dust closest to the explosion, where its light was brightest before it had much opportunity to spread through the surrounding space). This makes light echoes excellent probes of the dust between us and the supernova. Likewise, when we see blue starlight scattered off dust in such regions as the Pleiades star cluster, most of what we see is the dust on the front side – something our common sense is not tuned to by the large objects we encounter in everyday life.

We can also use light echoes from less violent events to probe some stars' immediate surroundings. In January, 2002, the star V838 Monocerotis suddenly brightened by a factor of 10,000, beginning a six-week set of changes in brightness and color before fading to its usual brightness. This combination of brightening and reddening did not fit the pattern of known kinds of variable stars, including typical nova outbursts triggered by the dumping of fresh hydrogen from a companion star into a dense white dwarf. Ground-based telescopes showed a small fuzzy path around the star soon afterward, appearing to expand so rapidly that it could only be a light echo, potentially telling us something about the sur-

---

[5] Some derivations refer to this form as a paraboloid, obtained only if we consider ourselves infinitely far from the star in question. Not that it makes a measurable difference for astronomical situations, but if we are looking at geometric definitions, we are allowed to be pedantic.

Fig. 2.2. Two ringlike light echoes from Supernova 1987A appear in this image obtained 9 months after the light from the explosion passed us. Both rings, marking where we see the supernova explosion illuminate foreground clouds of interstellar dust, are brighter toward the north side where there is more dust. Image data obtained with the 1.5-meter telescope of Cerro Tololo Inter-American Observatory, Chile.

roundings of this unique event. A spectacular set of *Hubble* images (obtained in a program led by Howard Bond) traced the expansion of the light echo over the ensuing two years, as the light illuminated dusty regions and cavities around the star (Fig. 2.3). Adding the time dimension breaks the flatness of the pictures; each dust patch tags itself with location along our line of sight. V838 Mon is surrounded by thick shells of dusty material centered on the star, telling us that it has had some kind of outburst before – thousands of years before. The expansion of the light echoes also revealed the distance – despite its appearance against the part of the Milky Way that contains one of the spiral arms closest to the Sun, V838 Mon floats in the isolation of the outermost disk of the Milky

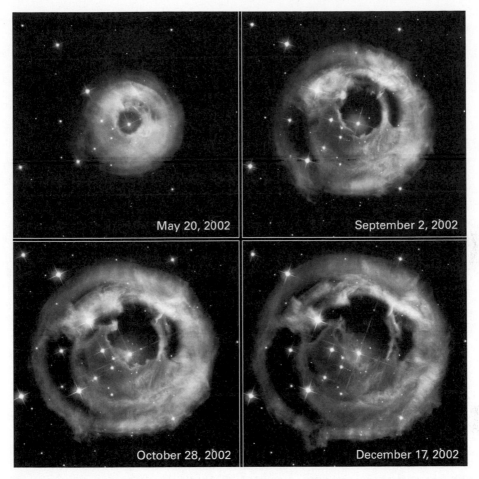

May 20, 2002

September 2, 2002

October 28, 2002

December 17, 2002

Fig. 2.3. This dramatic series of *Hubble* images captures the light echoes from the 2002 outburst of V838 Mon, as the flash illuminates dust in interstellar clouds progressively farther in front of the star. (NASA, ESA, Howard Bond, and the Hubble Heritage project.)

Way, 20,000 light years from us (and 46,000 from the Galactic Center). Such clues may lead us to understand what happened to this star, and how this fits into the overall pattern of the lives and deaths of the stars.

Less dramatic changes in light also produce light echoes, likewise useful in mapping the sources' neighborhood. The Cepheid variable star RS Puppis is surrounded by enough dusty material to have a permanent set of light echoes, new ones appearing in front of the star as older ones expand into invisibility. In this case, each of the light echoes brightens and dims following the 41-day pulsation period of the star, but with different phases depending on the light-travel time of each dust region from the star.

The opposite of a light echo (a "dark echo"?) can also show us the unseen. Some young stars are heavily shrouded in dust, so we may see their visible light only as it is scattered off of surrounding clouds. Some of these nebulae are remarkably variable, changing the brightness and shape of important features from year to year. One such object, NGC 2261, became known as Hubble's Variable Nebula, and was the target of the first official photograph taken with the Palomar telescope in 1948. Here again, the changes could not represent actual motion of gas or dust, because that would require motion beyond the speed of light. The key to unravelling the variations was noting that the same features always reappeared when certain parts brightened; it was the illumination that changed, not the nebula. We are seeing the shadows of dusty patches forming from material shot out by the mostly-hidden star R Monocerotis (Fig. 2.4), as they are cast outward through the nebula. Similarly, "dark echoes" appear in the dusty surroundings of UW Centauri. This is a so-called R CorBor star, one in which carbon compounds condense in its vast and cool atmosphere. When a cloud happens to condense right in front of such a star, it will all but vanish from our optical view until the sooty patch moves or thins. Careful study of the surroundings shows "searchlight beams" shining out between the absorbing patches, allowing us to trace what happens so close to the star that not even *Hubble*'s power will show any detail.

We can even follow light echoes too distant or too small to separate from the image of the illuminating object. Some supernovae in distant galaxies show a slight brightening well after the explosion, which has the right properties to be an echo from the light outburst scattering from dust close to the explosion. But they have, perhaps, been most informative in probing the structure of the brightest individual light sources in the Universe – quasars and the nuclei of Seyfert galaxies.[6] One of the most surprising observations surrounding the discovery of quasars went beyond the enormous power output implied by their brightness and distance. Routine brightness measurements, made in some cases when it was still thought that these might be some kind of peculiar stars in our galaxy, showed that quasars flicker in brightness from week to week (and in X-rays, we find that they may flicker measurably in only a few minutes). In itself, this shows that the light source cannot be much larger than our Solar System, despite outshining an entire galaxy. Except for some kind of cosmic conspiracy, no light source can be seen to change its brightness faster than the time it takes light to cross it front-to-back. For example, if someone were to switch off the Sun so that its entire surface blacked out at once, we would see the event spread out over about two seconds. First the center of the disk would vanish, with the region quickly expanding as the final light from more distant parts of the surface reached us, and finally from a rapidly thinning ring at the edge of the Sun's disk.

---

[6] See Chapter 7 for more about these objects and their role in tracing massive black holes.

Fig. 2.4. Hubble's Variable Nebula, NGC 2261 in Monoceros. The bright, but mostly dust-shrouded, star R Monocerotis illuminates the surrounding dusty region. As dense clumps of dust pass close to it, their shadows shoot outward and change the appearance of the nebula from year to year. Because of the finite speed of light, the shadows of moving objects do not point directly away from the star, and the appearance we see does not match the nebula's illumination at a single time in its frame of reference. (Image from the University of Alabama's 0.4-meter telescope.)

2.4 The message of light

Untangling what we know of quasars and their kin, like our knowledge of stars, relies heavily on spectroscopy – the analysis of radiation by how much we see arriving in very narrow bins of wavelength (or equivalently, frequency or energy). The power of spectroscopy originates in the connection between the structure of atoms and the kinds of radiation that they can emit or absorb. The frequency and wavelength of any kind of radiation are associated with the energy carried by each packet of the radiation. When it acts as a particle, such a packet is a *photon*. Identifying the photon was one of Einstein's contributions to quantum physics, the core of his work on the photoelectric effect which was principally cited in his Nobel Prize award. This effect is at the heart of solid-state cameras and solar electric power systems. When certain metals are illuminated by radiation above a certain threshold energy (or, equivalently, with wavelength shorter than a critical value), an electrical current is produced (or likewise equivalently, the metal atoms eject electrons). This posed a problem if we understood light purely as a wave phenomenon. The current starts to flow immediately when the light reaches the working material, but energy from a wave would be so widely spread among the constituent atoms that it would take a very long time to build up enough to liberate electrons. Einstein realized that the problem would be solved if radiation behaves like a stream of particles, at least while interacting with matter. Only those atoms which happen to interact with photons give off electrons, and they can start doing so immediately.

This notion, of radiation behaving in discrete chunks, fits with Max Planck's discovery only a few years earlier, that the distribution of radiation from a so-called blackbody[7] required that the energy of the radiating system (at least) could take on only certain discrete values (hence *quanta*). Planck had originally explored this notion almost accidentally, and Einstein's explanation played a key role in showing that indeed, radiation interacts with matter in these quantized units – photons. The energy of a photon is related to its frequency (by a value since designated Planck's constant, h).

There are several ways of measuring the frequency or wavelength of radiation, with the most familiar being glass prisms for visible light. Light is refracted as it enters and leaves the glass, by angles depending on how much the light slows down within the prism compared to its surroundings. Since this loss of speed depends on the wavelength (as well as the chemistry of the glass), the net result is that different wavelengths emerge in different directions. It was this phenomenon that led Isaac Newton to note that "white light" must be a composite of all colors, while later, more refined applications showed that various heated substances give off characteristic patterns of radiation, and that some of these patterns are evident in the spectrum of sunlight. Besides prisms, there are additional ways to decompose a beam of light (or any other kind of radiation) to tell how much is arriving at various wavelengths. Owing to the ability of light to act like waves (as well as particles), it passes a series of obstacles in different directions depending on wavelength. This can be used in the form of a *diffraction grating* – a surface, transparent or reflective, shaped into myriads of tiny parallel grooves. As everyday examples, the grooves on the surface of a CD or old-fashioned vinyl record will show rainbow hues when properly viewed, as the various wavelengths diffract at differing angles. The angular spread of the radiation depends on its wavelength compared to the spacing of the grooves (as well as whether it arrived perpendicular to the surface). At the highest energies, for hard X-rays, we cannot actually machine gratings with rulings fine enough to produce a detailed spectrum by diffraction, but nature steps in to help. There are crystals (of silicon, and lithium fluoride) in which the spacing between atoms is just right to act as the grating spacing. At the opposite end of the electromagnetic spectrum, we likewise use very different techniques for radio astronomy. There, a spectrometer may be a bank of electronics designed to correlate the signal

---

[7] A blackbody is a theoretical construct, which can be closely approached with careful laboratory experiments. It consists of a material which is a perfect emitter as well as absorber of radiation surrounding a cavity; in the laboratory, one has to allow a small hole for inspection of the radiation within. With the discovery of the cosmic microwave background, it became apparent that we live inside a vast blackbody cavity, with boundary conditions in time rather than space (Chapter 8).

with itself after applying multiple delays corresponding to various frequencies. But all these technological differences should not obscure the basic similarity, of measuring how much radiation arrives at different frequencies (or wavelengths), because the patterns revealed in this way are vastly informative as to how the radiation originated and what has happened to it along its path.

By the early twentieth century, physicists recognized that the spectral pattern of a given element was a window to the structure of its atoms. The key was in combining the conservation of energy (for these processes, the "mass" part of mass–energy makes very little difference) with the new principles of quantum physics. The classical theoretical framework of electrodynamics set out by James Clerk Maxwell a generation earlier showed how electrons could circle an atomic nucleus, once we knew that electrons are negatively charged while the nuclei carry positive electrical charges. However, there was a contradiction in the predictions of this theory. Radiation is given off whenever electrically charged particles accelerate, and electrons under the electrodynamic attraction of a nucleus are under constant acceleration. As an electron radiates energy, that energy is lost to its motion, so it would spiral quickly into the atomic nucleus. The very fact that our atoms continue to exist tells us that this understanding of atoms is seriously incomplete.

Explaining the structure and stability of atoms led to the "first quantum theory", a sort of intellectual way station for exploring quantization of energy and the interaction of particles and radiation. It was in this context that Niels Bohr proposed the 'Bohr atom', the model so familiar from schoolbook diagrams, in which the tiny nucleus is orbited by whizzing electrons. Shortly before meeting Bohr in 1911, Ernest Rutherford had found that most of the mass of an atom resides in a tiny nucleus, and their exchange of ideas played a key role in developing (and eventually superseding) this early theory with the later quantum mechanics. In this model, Bohr postulated that electrons could orbit only in paths for which the angular momentum (in classical physics, given by the product of the electron's mass, velocity, and orbital radius) had a value that was an integer multiple of a fundamental constant. Furthermore, and finally connecting the structure of atoms to light, an electron could emit or absorb a quantum of radiation (a photon) if that quantum carried enough energy to enable the electron to make an exact transition into another allowable orbit. This theory was first applied to the simplest atom, hydrogen, and accounted for the regular arithmetical pattern seen in the spectral lines of hydrogen (and, in fact, all elements when stripped down to a single electron).

Thus, in a particular kind of atom, only those electron orbits exist which satisfy a quantum rule so that the angular momentum of the orbit has one of the allowed values. The situation is more complicated, and fuzzier, when we include the refinements that constituted a sort of "Second Quantum Theory" (which became today's quantum physics). Rather than following a particular orbit like planets around the Sun, we must think of electrons as existing in a probabilistic cloud around the nucleus, where each one has a particular probability of being in various places if we make the atom interact in such a way as to localize an electron. Different "orbits" correspond to different shapes and sizes for these probability distributions (known as the electrons' wave functions). However, we can understand much of what we see in astronomical spectra by using the "cartoon physics" in which we can sketch electron orbits, always keeping in mind that if we examine the situation in detail this kind of visualization will eventually prove inadequate (and downright misleading).

For any kind of atom, there exists in principle an infinite number of possible electron orbits, with progressively more energy (approaching the limit in which the electron has so much energy that it flies away from the nucleus completely). If there is an unoccupied orbit with less energy, the electron will naturally decay to that lower energy level, radiating the energy difference away as a photon carrying that amount of energy. The time it takes to undergo this decay cannot be predicted individually; this is one of the purely statistical features of the quantum world. However, we can determine the half-life or rate of decay for a large sample, which depends on how much the wave functions of the two orbital states overlap. The more the two states look alike, the faster the decay takes place. Conversely, an electron in some low-energy orbit can absorb a photon if its energy is a sufficiently close match exactly to the energy difference needed to push the electron into a possible higher-energy orbit (or enough energy to launch the electron free of the atom, a process known as *photoionization*). The spectrum of photon energies seen from, for example, excited hydrogen atoms, tells us the differences in energy among the possible electron orbits (Fig. 2.5). This realization set off years of research in which the study of atomic structure was the study of spectra of various elements in the laboratory, under whatever conditions of temperature and pressure could be attained, over the widest span of wavelength possible. Fitting the results together could turn into a logic puzzle. What combinations of spectral lines could add up in energy to give consistent sets of orbits, and how did they mesh with what we were coming to expect in the light of quantum-mechanical calculations that showed which orbits would have the most likely transitions?

The quantum understanding of atoms gave us keys to unlock the Universe. We could work out what spectral lines should be emitted or absorbed, and how strongly, by a gas of specified composition, temperature, and pressure. Most famously, this showed the way to measure the chemical makeup of the stars, beginning only a few years after Auguste Comte wrote, "we shall never by any means be able to study their chemical composition". But less than a century later, astrophysicists were unravelling the composition of the Sun and beginning to compare it to other stars, and by now we are discussing the chemical history of the stars in galaxies when the Universe was only a tenth of its present age, and uncovering the history of the production of the chemical elements. Light is wonderful, and carries unexpected riches.

Some of the basics of why astronomical spectra appear as they do fell into place with three laws distilled from laboratory experience by Gustav Kirchhoff (inventor of the spectrograph in 1859, working with Robert Bunsen of burner fame). Kirchhoff was an unusually prolific observer of patterns in physics. His three laws of spectroscopy should not be confused with Kirchhoff's laws of electric circuits, or Kirchhoff's laws in thermodynamics. His laws of spectroscopy tell

Spectral Lines   **Atomic Energy Levels**

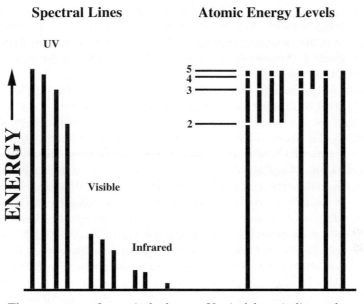

Fig. 2.5. The spectrum of atomic hydrogen. Vertical bars indicate the amount of energy in each of the spectral lines given off or absorbed by hydrogen, restricted for simplicity to those among the first five possible energy levels. At left they are sorted by wavelength. At the right, they are reorganized and duplicated to show how they fit together to make the "ladder" of energy states telling us the structure of the hydrogen atom. A similar logic puzzle, albeit more complicated when many electrons are involved, exists for each element and ionization state.

us that, first, a hot and dense gas produces a continuous spectrum – a smooth distribution of energy with no narrow peaks or dips. Second, a hot gas at low density produces an emission spectrum as long as we see it against a cooler background. And third, a cool gas seen in front of a source of continuous background light produces an absorption-line spectrum. Our understanding of how to derive the makeup of stars can be traced to Kirchhoff's observation that the wavelengths of a pair of dark lines in the yellow–orange part of the solar spectrum exactly match the bright lines seen in a sodium flame in the laboratory. Applied to astronomical spectroscopy, this tells us that a gaseous nebula, when heated by shock waves or starlight, will produce an emission spectrum. Most stars, with intensely hot and dense interiors surrounded by relatively cooler layers of atmospheric gas, will show absorption spectra like the Sun's.

Now the pieces fall into place for analyzing the spectra of objects in deep space. The relative strengths of certain emission lines (sets from oxygen and nitrogen are particularly useful) reveal the temperature of the gas. Knowing that, we can determine the effectiveness of various kinds of atoms (including ionized species without their full complement of electrons) in producing various emission lines, and thus the number of atoms of each kind. Of course, the greater the accuracy we want, the more complications enter. We seldom see homogeneous gas masses at a single temperature and density, so what we can observe is a mix of different regions. And some kinds of atoms can be especially hard to see – carbon is important in tracing the history of element production, but all its strong spectral features occur deep in the ultraviolet or far in the infrared. Uranium can be used in a few stars for radioactive dating, but only in stars which are poor in carbon, because the strongest uranium feature is overlapped by a weak line from (much more abundant) carbon.

Decoding a star's spectrum in detail requires numerical modelling, essentially asking which of a library of synthetic results is the best match to the real star. One takes a particular temperature and surface gravity for the interior of the star, and generates sets of atmospheres to wrap it in, each with a different composition and possibly vertical structure. The vertical structure will be constrained by the properties of gas at the appropriate temperature, so changing the chemical mix will change the local temperature and hence the way spectral lines form. Fig. 2.6 compares the emission and absorption spectra of a nebula and a star.

Spectra rich in distinct emission or absorption lines are correspondingly rich in information. It is spectroscopy which has delivered most of what we know about the Universe beyond the reach of our space probes, and spectroscopes which are in use for most of the observations with large telescopes. Spectra with distinctive patterns of lines are also custom-made for measuring the wavelength shifts that mark relative motion, effects of gravitation, and the stretching of spacetime on a cosmic scale.

2.5 Echoes in quasars

Spectroscopic analysis of quasars and their kin shows that the central engine is surrounded by glowing gas, either shining by absorption of radiation from the core or some fast reprocessing of that radiation. The Doppler-shift spreads and amount of energy needed to excite various kinds of atoms we see in the spectra show that some of these spectral features come from gas very close to the core, and others from many light-years away. Starting with the variations we see in the central source, we can use the same kind of analysis used to derive the appearance of light echoes around exploding stars to ask how far from the core, and in what kind of form, we find the components of surrounding gas. In the context of galactic nuclei, this has come to be known as reverberation mapping. Its beginnings date to the chance observation of changes in the spectra of a handful of quasars and Seyfert-galaxy nuclei observed several years apart. Emission lines could be stronger or weaker relative to the blue continuum light that is characteristic of these objects (perhaps a result of matter swirling around the central black hole; Chapter 7), and in some cases the very broad parts of spectral lines that betrayed Doppler shifts from the most rapidly-moving gas would brighten and fade, appear and disappear. Fig. 2.7 illustrates the broad spectral lines seen in quasars.

Perhaps the first systematic project attempting to exploit these changes to answer major questions about active galactic nuclei (AGN) were carried out in

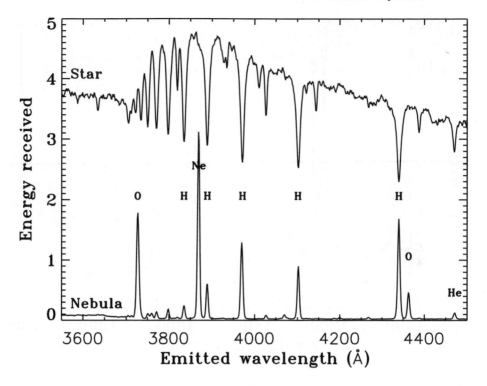

Fig. 2.6. The two major kinds of astronomical spectra are illustrated by observations of a star and a nebula. The upper line shows the spectrum of the hot B-type star Hiltner 600, in which we see absorption lines (especially due to hydrogen) as the continuous radiation from the interior passes through the cooler atmosphere of the star. The lower line shows the spectrum of the nebula Jonckheere 900, whose gas is excited by a very hot central star. The gas produces emission lines; those from hydrogen match the wavelengths of absorption in the stellar spectrum. Other prominent emission lines arise from ionized states of oxygen, sulfur, and neon. The hydrogen lines in both objects show a regular progression in intensity, becoming weaker to shorter wavelengths (left). This sequence appears to be broken in the nebula by the line near 3970 Å , because of blending with a neon line of almost the same wavelength. (Data from the 2.1-meter telescope of Kitt Peak National Observatory.)

1978–81. A study of the ordinary visible-light range was carried out by Robert "Ski" Antonucci and Ross Cohen while both were graduate students, working at Lick Observatory. Using a spectral scanner on the then–new 1-meter telescope, they were able to observe the bright Seyfert galaxy NGC 4151 thirty times within a 14-month period, with only a single gap longer than one month when this galaxy lay nearly behind the Sun. Their work followed a space-based study carried out by an international team using the *International Ultraviolet Explorer* (*IUE*) (Fig. 2.8), who had compared seven observations of the ultravio-

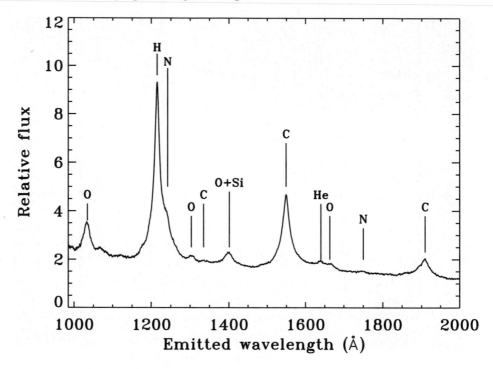

Fig. 2.7. An average quasar spectrum, showing the ultraviolet range which is seen from satellites for low-redshift objects and, shifted into visible wavelengths, also by ground-based instruments at high redshift. Emission lines which occur in gas at high densities are very broad, sometimes blended together, as a result of rapid internal motions in the gas. The carbon lines at 1549 and 1909 Å are especially noteworthy, occurring at higher gas densities than seen in normal nebulae. These are some of the spectral lines monitored for changes following fluctuations in the intensity of the continuous light, which is thought to trace the deeper-ultraviolet ionizing radiation which most directly powers the gaseous emission, and arise close to the central black hole. Data from *Hubble* observations, provided by Wei Zheng.

let spectrum of NGC 4151's nucleus taken over an 11-month period. *IUE* marked the coming-of-age of space astronomy. Guest observers would go to control centers in Spain or Maryland and interact with telescope operators in a manner similar to that of large ground-based telescopes, and introduced a generation of astronomers to the use of spacecraft. Constructed jointly by US, UK, and European space agencies, it was launched in 1978 for a three-year nominal mission, carrying a 45-cm telescope and ultraviolet-sensitive TV detectors behind spectrographs designed to slice the ultraviolet range in four different ways. It was finally turned off in 1996, having sextupled the initially expected lifetime and delivered well over 100,000 observations of all kinds of celestial objects as seen in the ultraviolet.

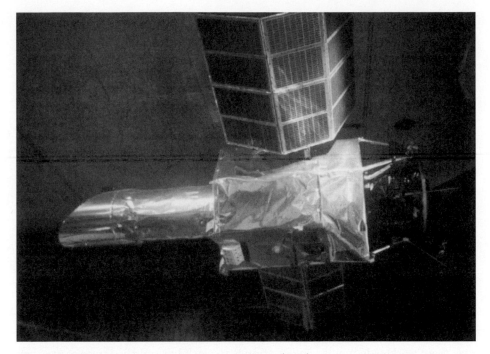

Fig. 2.8. The *International Ultraviolet Explorer* (IUE), represented by its engineering mockup in Washington's National Air and Space Museum. The telescope tube contains a mirror of 45 cm aperture. The original was launched into an elliptical geosynchronous orbit in 1978, and acquired more than 100,000 ultraviolet spectra during almost 18 years of operation.

The *IUE* data showed that when the central radiation source faded, so did the emission from surrounding gas, but the intervals between observations were too long to say how quickly this happened. Probing the time lag between the central source and surrounding gas could lead to important advances in our understanding of active galaxies, including the mass of the central objects long thought to be extremely massive black holes. One way to estimate the mass of the black holes was to take the widths of the broad emission lines in their spectra as Doppler shifts, assume a typical distance of this gas from the center, and hark back to Kepler's laws to ask what mass was needed to hold this gas in orbit. The spectral lines gave only the merest clues to the size of the region; it was too small for telescopes to resolve, and some parts of the gas had densities falling in interesting ranges from a few hundred to hundreds of millions of particles per cubic centimeter.

Antonucci and Cohen's results showed promise, and by then a large following was developing for such monitoring projects. They could show that the gas responds to the brightness of the central source, that different components respond with different time delays (and thus have different sizes ranging up to a light-year), and that essentially the same configuration of surrounding gas is present all the time, whether brightly or dimly illuminated. More extensive observations might be able to resolve the long-standing question of what the surrounding gas was doing – coming, going, or just milling about. What we see in these projects is exactly like light echoes from exploding stars, except that we have to infer the echo from spectral features because it is too far away to make out details in an image. In analyzing the results, we have to fold in the expanding paraboloid of locations where we see gas that is, for example, just now responding to some brightening or fading of the central object. And even though we cannot make out any structural details, a spectroscopic approach tells us something else immensely useful – which spectral lines, and even which parts of the spectral lines, respond first or last.[8] So, if the gas were mostly flowing outward from the center, the near-side gas will give off blueshifted spectral lines, which we would see respond first to a change in the illumination from the center (because the nearside gas is seen by a more direct light path). Conversely, for inflowing gas, we would see the redshifted part of the line change first. In reality, most of the gas does not show a net inward or outward motion – it consists of material orbiting in the local gravity, with any net radial motions confined to some small fraction of the material.

But seeing this kind of detail would take yet more expansive projects, far beyond the means of a single research group. At international meetings in 1987 – in Segovia, Spain, and Atlanta – the idea took root that astronomers should organize truly worldwide monitoring programs, to work beyond the limits of weather, observing endurance – and the fact that other astronomers needed the telescopes too. Coordination was needed not only between numerous astronomers working at observatories worldwide, but with the operational constraints of spacecraft. With its unique access to the ultraviolet spectrum, results from *IUE* would be central to such a campaign. To have the best chance of sampling telltale changes, *IUE* would need to observe a single object for sixty 4-hour sessions per year. Along with the demands of data analysis and breadth of scientific interest in the program, a large number of co-investigators was important in demonstrating to the allocation committees the importance and unprecedented scope of the project. Fortunately, the galaxy did behave well, showing variations in the continuum from the central source and the emission lines from gas at various distances (Fig. 2.9). The rationale for repeated observations of active nuclei was so compelling that even in the early years of *Hubble*, when allocation panels would argue over every 15 minutes of requested time, the International AGN Watch team – grown to over 130 co-proposers – managed to get a precious forty orbits of time to follow changes in the spectrum of the Seyfert galaxy NGC 5548. These results not only specified the sizes of some of the emitting zones, but gave strong hints of their shapes. By this time, the link to light echoes had been implicitly recognized with the introduction of a so-called transfer function that

---

[8] The analogy to light echoes is correct as long as the gas responds quickly to changes in the radiation environment, which is a good description for most of the gas around active galactic nuclei. If it took the gas a year to change its properties after the nucleus flared, the analysis would be much less sensitive.

relates changes in the illuminating source and gas clouds, which explicitly folds in the time delay between our lines of sight to the two and ellipsoidal structure of the lines of constant time delay.

Several additional active nuclei were eventually observed in similar campaigns, with Brad Peterson at Ohio State University providing an organizational anchor. This proved to be a powerful use of *IUE* in its final few years, when most bright active-galaxy targets had been observed at least once and the new wrinkle of variability was important. Furthermore, as it became more limited in flexibility of maneuver and pointing, the ability to "park" on a few targets for long times made such observations attractive. Eventually, the combined data could show that brighter nuclei have larger emission regions, and that the inner "broad-line" regions were smaller and more responsive than back-of-the-envelope calculations had once suggested.

2.6 What we see and what we get

It is not only light that can travel fast enough to make things appear quite different from their "instantaneous" structure. Jets of matter and wave motions can do it as well, once they reach a significant fraction of the speed of light.

In principle, light-travel delays affect the appearance of all manner of celestial objects. We see the far side of the Andromeda galaxy as it was about 100,000 years before our view of the front. But the galaxy takes about a quarter billion years to rotate, so nothing much changes in the big picture in a mere 100,000 years. Stars out in its spiral arms travel only about $0°\!.2$ in their orbits – not enough to disturb its appearance. Things would be very different if Andromeda were spinning near the speed of light. We would see its spiral form quite distorted and compressed on one side.

Some effects are more complex and start to break our intuitive expectations in the ways we have come to expect from relativity. In the astronomical arena, we can find objects moving so fast that we see their clocks run fast or slow (depending on whether that motion is toward us or away). We can see illusions produced by objects moving so close to the speed of light that they almost catch up with their own radiation, producing an optical illusion that would have caught Einstein's attention. These illusions can make streamers of material flowing outward from compact objects look as if they are moving in a different direction, and make their internal patterns seem to reverse. And these effects are encountered over and over, from black holes in our own Galaxy to the powerhouses at the center of quasars.

We do expect to see major distortions of the way we see material around the most highly relativistic objects, black holes, But to understand how strange these distortions would be, we need to introduce two new results of relativity. One, known as Doppler boosting or Doppler favoritism, belongs here. The other, the deflection of a light path by gravity, will be taken up in Chapter 4.

Special relativity deals, above all, with how measurements of physical systems transform between different frames of reference (since we can interpret

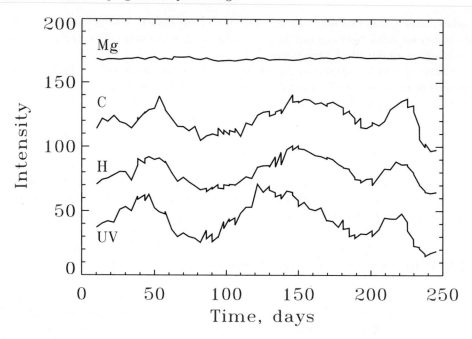

Fig. 2.9. These data, from the *IUE* campaign carried out by the International AGN Watch, show the time lag between the ultraviolet continuum in the Seyfert galaxy NGC 5548 and various emission lines. The UV continuum is believed to arise in a small region very close to the central black hole, while the emission lines come from cool gas much farther out. This distance, and some features of the gas distributions, can be reconstructed from analysis of how much of the gas responds to changes in the central source after various time lags. The hydrogen emission responds first, followed by the carbon emission a few days later. The magnesium emission region is much larger, perhaps a light-year in radius, and barely responds at all to these brief changes in the illuminating source. The data for different spectral features have been arbitrarily scaled for clarity. (Data from the International AGN Watch.)

their appearance only relative to a particular means of measurement). These transformations act on both the ways we measure space (sizes, locations) and time (speeds, frequencies). Their mathematical forms had been derived before Einstein, but interpreted as results of interactions with a medium – the ether – which permeated space and was the material in which light moved as a wave. In fact, the central transforming equations of special relativity are sometimes known as the Lorentz transforms, and the scaling factor by which energy increases with velocity is the Lorentz factor, in honor of the eminent Dutch physicist Hendrik Lorentz (1853–1928). He was among the first to realize that James Clerk Maxwell's theory uniting electricity and magnetism was not just a masterly piece of theoretical development, but one ripe with further directions and connections to the rest of physics. He presented his wrestling with the old physics and first glimpses of the new in an 1895 book, *Versuch einer Theorie der Elektrischen und Optischen Erscheinungen in Bewegten Körpern* (*Toward a Theory of Electromagnetic and Optical Phenomena in Moving Bodies*), whose title is worth quoting because of the echo it found in Einstein's 1905 announcement of special relativity. Our library at the University of Alabama has a copy of the 1906 reprinted edition, whose timing ended up marking an unpredictably poor publishing decision (which may also explain why no English translation seems to have been published).

Lorentz's book marks a field in transition. He was aware that there must be tension among various well-supported ideas – ether, Maxwell's fabulously powerful mathematical theory of electromagnetism, Fizeau's earlier and intellectually attractive theory of light propagation, experiments failing to find any experimental evidence of the ether. What may be most telling is that Lorentz derives the precise form of the apparent contraction in length for a moving object we see (a factor known to this day as the Lorentz factor, the quantity most often appearing when applying special relativity in an astronomical context), but cannot get out of his head the idea that light must be a vibratory motion in some physical substance, the luminiferous ether (ether for short). Lorentz worries about connecting the electromagnetic theories to the newly-supported ideas of electrons and charged atoms (ions), a quandary which finds resonance at the start of the twenty–first century in the steadfast refusal of relativity and quantum mechanics to play together nicely in the mathematical yard. He recognizes that preserving the action of physical laws (such as electromagnetism) without self-contradiction means that something has to be dispensed with in our intuitive notions of what is constant among objects. Still, it took Einstein's work to take the necessary conceptual leap and eliminate any measurable role for "absolute rest".

The Lorentz factor enters into such diverse phenomena as how much we see time in some moving object run slower than ours, how much brighter or fainter a rapidly moving object appears than it would be at rest, and the apparent speed of moving objects because of the transformations of space and time announced in special relativity.[9] In brief, if we see an object moving with a particular velocity relative to us, in our frame of reference, we will see its time frame run slow by the Lorentz factor.

With George FitzGerald in Ireland, Lorentz found that we would measure an object in motion (with respect to us) to be shortened along the direction of this motion – the Lorentz–FitzGerald contraction. At first it was thought that this must be a physical contraction produced by the ether, but Einstein, in doing away with the apparent need for such a medium, showed that the

---

[9] Its mathematical form is $1/\sqrt{(1 - v^2/c^2)}$, which shows that we need to deal with motions as rapid as 15% of the speed of light, 45,000 km/s, in order for these effects to show a 1% departure from what we would expect from the old-fashioned Newtonian rules. The fastest of our spacecraft, the *Galileo* probe falling into Jupiter's atmosphere, reached a blistering 47 km/s, or 0.015% of c. The Universe furnishes us with many places to see much larger differences.

contraction results simply from the transformation between reference frames. Lorentz himself grasped this from Einstein's work, one of his final publications being "The Einstein Theory of Relativity: a Concise Statement". Applying this symmetric principle to a favorite science-fiction scene, we would see an alien starship contracted, and they would see ours similarly distorted. Not knowing the "actual" shapes of some of the things which we see in space moving at such velocities, this effect by itself is not one that astronomers have had much occasion to see.

However, Lorentz's results also showed that we would see an object passing us at high velocity appear to be rotated from its "actual" orientation, simply because light from the near edge reaches us sooner than from the far edge, and hence we see the near edge as it was slightly later and farther along in the object's motion (Fig. 2.10). This apparent rotation may play a role in how we see the numerous jets of energetic particles seen flying away from energetic objects throughout the Universe – from neutron stars in the Milky Way to the gigantic black holes in distant quasars. [10].

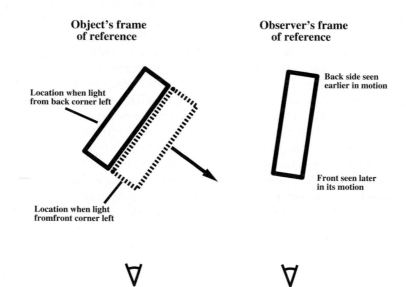

Fig. 2.10. The Penrose–Terrell rotation, observed when we look at an object moving relativistically in our reference frame. Light from its near side takes less time to reach us, so we see it at a later time and at a later location along its path than we see the far side. The net effect is an apparent rotation of the object. This phenomenon enters into the structures we see in quasar and radio galaxy jets; they are actually at smaller angles to our line of sight than the observed structures would naively suggest.

This effect, which almost acts like a rotation except for not letting us see the actual back side of the moving object, has helped to confuse astrophysicists

[10] This is sometimes called Penrose–Terrell rotation, after the people who worked out the gritty details of who sees what and why.

about features seen in the narrow jets emerging from some extreme objects – black holes and possibly neutron stars (which we will return to in Chapters 3 and 7). The first of these jets was discovered on photographs with the 0.9-meter (36–inch) Crossley telescope of Lick Observatory as long ago as 1918, in which the bright galaxy Messier 87 in the Virgo Cluster shows what H.D. Curtis described as "a curious straight ray" pointed from the core. With the discoveries of radio galaxies and quasars in the 1950s and 1960s, M87 was joined by hundreds of other galaxies and quasars in showing such features when observed by radio telescope, now being distinguished by its proximity and the fact that its jet is by far the brightest emitter of visible light among all these instances. As we became able to see their structure more clearly (mostly due to the introduction of multi-telescope interferometry in radio astronomy), their common patterns became clear. Many of these may be seen in the excellent *Hubble* image shown in Fig. 2.11.

Jets invariably trace back to a core – either the nucleus of a galaxy or a brilliant quasar. They are (by definition) narrow, in some cases thirty times as long as they appear wide. Often (but not always), they arise in a core showing the brilliant pinpoint of an active nucleus (although there are some radio galaxies with radio jets but a calm, visually unremarkable galaxy core at their root). Jets may continue outward in nearly straight paths for several million light-years, which indicates that they must last at least the several million years that it would take them to grow so long (spraying outwards at relativistic speeds). Their linearity also shows that the central source has some preferred direction to eject matter, a direction which stays the same over long spans. As we shall see in Chapter 7, the prime suspect for launching these jets is a supermassive black hole and surrounding disk of matter.

Finally, we can tell that the jets are indeed moving outward, and doing so near the speed of light. Careful measurements of various blobby features in jets shows that, in many cases, we can measure their outward movement, and that these motions are often fast enough to suggest a flow speed close to the speed of light. Indeed, in some cases the speeds we measure in this way are *too* fast, working out to several times the speed of light. These are the *superluminal sources* (section 2.6).

Jets are ubiquitous across the sky, once we have radio telescopes capable of picking out enough detail. Thousands of galaxies and quasars show them, with sizes ranging from a few hundred light-years to several million. And, in a surprise that is still under active discussion, the *Chandra* X-ray observatory is seeing many more jets than would have been expected, showing that some new process enters when we look at such high-energy radiation. The front runner for now seems to be interaction between the relativistic particles in the jets and surrounding radiation, especially the cosmic microwave background.

These jets present only two outstanding problems – how they are initiated and how they maintain their identity so long. Calculations have shown that, quite broadly, magnetic fields in material swirling at the high speeds encountered near a black hole can launch charged particles at high speeds in opposite directions. Since the escape velocity so deep in the black hole's vicinity is close to the speed of light, any material that escapes to large distances will have started out that fast, so it would not be a surprise to find that it still has enough energy to be relativistic until it runs into something that would slow it down. Even at these high velocities, there is still something to be understood about the length of jets. If the central engine were "merely" spraying material outward in a narrow beam at, say, 99% of the speed of light, that beam would broaden rapidly due to the random internal motions of the particles, unless they were launched quite accurately parallel and avoided interacting with one another on the way. Again, magnetic fields may be the key, with the field becoming stretched and wrapped in such a way as to hold the charged particles in such a narrow stream.

Fig. 2.11. The jet and core of the nearby radio galaxy M87, in a *Hubble* image. The reddish starlight is strongest near the galaxy's core, while the bluer light of the jet shows delicate twists and planar brightenings to a projected distance of about 6,000 light-years from its source. (NASA and the Hubble Heritage Project.)

Jets furnish particularly powerful demonstrations of the ways high speeds can alter our view. The nearby jet in M87 – the one discovered back in 1918 – has been a Rosetta stone for these objects, because it is nearby and observable in great detail not only in the radio domain, but also in the optical, ultraviolet, and X-ray regimes. This jet shows intriguing internal structures, which look much like the internal shock fronts that should build up if the original flow changes in speed or density (Fig. 2.11). Such shock waves would run perpendicular to the axis of the jet's motion, and appear like rungs on a ladder in numerical simulations. The possible shock features in M87's jet look very narrow, as if we see them (and the whole jet) almost from the side. But this conflicts with the speeds of motion of some of the knots in this jet, which can be measured not only by comparing radio maps taken years apart, but from *Hubble* images over a span of several years. If the jet is moving outward at 90° to our direction of sight, we should not see the speed of its knots reach such high velocities. This is one of the paradoxes of relativity – the transverse Doppler shift, caused as "moving clocks run slow", overcomes the sheer speed of an object unless it is approaching us. Reconciling these two estimates of the jet's direction brings us back to something that was implicit in the work of Lorentz – the apparent rotation of a rapidly moving object (that is, the Penrose–Terrell rotation). At its base, this phenomenon will occur because we see the front of a moving object by light that left it later, having less distance to travel to reach us, and the far side at a earlier time (when it had not moved outward as far from the galaxy core). Applied to M87, this means that the jet's internal structure will appear nearly sideways to us for a surprisingly large range of actual pointing angles, and could even be pointing within 15° of us and still show a shock front almost edge-on.

These travel-time effects can change a jet's appearance in even more remarkable ways. Instabilities in such plasmas can give rise to helical twists and filaments propagating outward along the jet, and we see some nearby jets showing such filaments winding their way outward. But, the twists we see may not be the ones in the jets. These same time delays can reverse the apparent direction of a helix, depending on how fast the pattern moves and our viewing angle. The pattern does not necessarily move outward at the same speed as the particles in the jet, just as the wave patterns in water do not move at the same speed as the water molecules (or more visibly, corks floating in the water).

The Lorentz factor enters once more in our analysis of these astrophysical jets, changing not just how we see their structures but how we see their time pass. This translates into our seeing approaching jets much brighter than we would if their matter stood still (with respect to us, as always), and the receding ones appearing dramatically dimmer (often to the point of invisibility). Einstein showed that not only our measurements of the spatial dimensions change with motion, but the time dimension as well (eventually breaking Newton's notion of "absolute, true, and mathematical time, of itself and from its own nature"). To our instruments, the time on some moving object will be changed in two ways. First, because of this relative motion, we will see time pass more slowly on the moving object, in a ratio given by the Lorentz factor. On top of this, if the object is moving toward or away from us, the relativistic version of the Doppler shift will apply not only to the frequency of electromagnetic radiation, but to the frequency of any event – that is, to any kind of change that marks the passage of time. One result of this is Doppler boosting – if the object we observe is approaching us, not only the frequency of the radiation is shifted, but the object will appear brighter, because the number of photons we receive in each second of our time is also increased, because each second of our time maps to many seconds in the object's own reference frame. This means that radio, X-ray and gamma-ray skies are full of things which appear bright only because they happen to be throwing material very close to the speed of light and very close to our direction, and that we may have to look very hard to see their counterparts pointed in some other direction.

2.7 Faster than light?

Much of the history of radio astronomy has been a quest for ever higher resolution. The long wavelengths in this regime mean that even large dish antennas deliver a view of the sky much fuzzier than we are accustomed to in visible light. Even the vast 300-meter Arecibo dish delivers images in the 11-cm band no sharper than does a human eye with 20/20 vision. Not only does this make it hard to distinguish structures in the radio emission, but it sets strict limits on how well we can match radio sources and their features with what we see at shorter wavelengths. The same wave nature of light which sets this limit also suggests a way around it.

When behaving as a wave phenomenon, radiation interferes with itself just as do more familiar water or atmospheric (sound) waves. We can use this fact to combine detections from multiple locations to produce much of the same information that would come from a single gigantic antenna stretching between the locations. Even simpler, the first application required only a single receiving instrument, looking from a cliff across a marine horizon to make a sea interferometer. As a radio source rises or sets over the ocean, some of the waves reaching the ocean's surface will be reflected back up into the receiver, and will interfere with the direct waves in a way that depends on the size and exact location of the radio source. Sea interferometers were employed for several years after World War II, mostly by the pioneering Australian group of radio astronomers, before being superseded by the more flexible possibilities of multi-element systems. Also in the late 1940s, several radio astronomers – particularly Martin Ryle (later Sir Martin) at Cambridge – constructed successively more elaborate interferometric systems combining the signals from two or more antennas, separated by growing distances. These could generate some important information about radio sources – location, size, whether one or two peaks were present – down to a resolution limited by the wavelength of observation and the antenna separation. The farther apart the antennas are located, the finer the resolution of the system, just as larger optical telescopes yield sharper images when our atmosphere is not a factor. For each pair of antennas, whether their responses are in phase or out of phase, and by how much, depends on how different the signal train is, which depends on the location and size of the radio source compared to a single sine-wave pattern on the sky. Multiple pairs of antennas compare the radio source to multiple sine waves, potentially building up a complete picture.

In 1960 Ryle took a crucial step with the invention of aperture synthesis[11], which has since been the tool of choice for most radio astronomy. In this technique, multiple antennas are used, but now some of them are moveable, and many observations at different times of day can be taken. This means that the source is compared to a great many sine waves, differing in direction on the sky as well as in spacing. With sufficient observations and enough different baselines,

---

[11] Ryle won a share of the 1974 Nobel prize in physics for this technique (Fig. 2.12), which it is still amazing to see in operation. Sure, the mathematics works out, but it is the sort of thing that does not seem as if it ought to work in practice. This award has been overshadowed in astronomical lore by the share of the same year's prize awarded to Anthony Hewish. Hewish's Nobel citation mentions his "decisive role in the discovery of pulsars". This seems to have been a *faux pas* on the part of the Nobel committee, since many contemporaries felt this unfairly slighted Jocelyn Bell's more immediate role (which was in fact recognized in Hewish's prize lecture). The Nobel organization seems to have learned to avoid such issues. The next Nobel Prize for work on pulsars included Joseph Taylor and Russell Hulse, who had been Taylor's student during the discovery of the binary pulsar PSR 1913+16.

a high-fidelity radio image can be built up (although many of us still have an archaic tendency to call them "maps"). This discovery led to ever more capable radio arrays – above all Westerbork in the eastern Netherlands, and the Very Large Array (VLA) in New Mexico. The VLA can resolve structures as small as a tenth of an arcsecond at its highest frequencies, nearly equalling *Hubble's* resolution for any kind of object that emits enough radio energy.

Fig. 2.12. In 1987, Sweden Post issued a set of stamps illustrating astrophysics in the Nobel Prizes. These two commemorate the awards to Ryle and Hewish, and depict both radio interferometry and the discovery of pulsars. (Copyright Sweden Post.)

These instruments opened up new phenomena to our eyes, with suitable assistance from the digital intermediaries that show us the results. Hundreds of galaxies and quasars displayed outflowing jets or radio lobes, suitable frequency tuning revealed the patterns of cold gas in galaxies, and we could peer within and through clouds of dust to see events in the center of our galaxy and the nurseries of starbirth. But astronomers could not yet penetrate the tiny cores of quasars, which remained beyond the limits of even these large arrays and were sure to contain fascinating phenomena.

The limiting size of these arrays was set by the need to physically transmit signals between the antennas for correlation, without either a timing interruption or unacceptable loss of strength. Beyond paths of about 40 km, wiring is not up to the job. For separations somewhat greater than this, carefully engineered radio links were up to the task. Astronomers at Jodrell Bank led the way here, in what eventually became the Multi-Element Radio-Linked Interferometer (MERLIN).[12] The curvature of Earth's surface, if nothing else, prevented more distant antennas from correlating their signals directly. Two enabling technologies remedied this – high-speed tape recorders and atomic-clock frequency standards. These allowed the two data streams to be captured and compared, albeit with a (usually) long hunt for the initial offset between time systems. Groups in the US and Canada swapped the technical lead several times in early 1967, finally leading to correlated detections ("fringes") between independent telescopes by May 1967, with both groups rushing to present their results at the same international meeting that month. This success cut the cord between

---

[12] Originally the Multi-Telescope Radio-Linked Interferometer (MTRLI), for reasons said to relate to Sir Bernard Lovell's distaste for the cuteness of the name. MERLIN was formally adopted only after his retirement as Director of Jodrell Bank.

interferometer elements, and consequently an interferometer could be composed of any set of radio telescopes anywhere on Earth (and eventually, off Earth as well) that could see the source at the same time.

With their tiny, brilliant cores, quasars were natural targets for this technique, known as Very Long Baseline Interferometry (VLBI). Many quasars (and radio galaxies) yielded structure to the new VLBI systems, while many of their cores remain too small for us to have measured to this day. Surely the most astounding of these early results came from repeated observations of the quasars 3C 273 and 3C 279 in 1971. A recent series of observations of 3C 279 is shown in Fig. 2.13. Such data series showed clumps or knots moving outward along the direction of the larger radio jets, but so fast that the velocity translated to several times the speed of light (assuming we had the quasar distances correct). Such *superluminal motions*[13] posed at the least an interesting puzzle, and possibly a major challenge to the edifice of special relativity, which had been so successful in so many arenas.

A wide range of explanations was proposed for superluminal motion. Perhaps there were various bright spots constantly appearing and disappearing, so the limitations of our data led us to infer motions where none existed. Or perhaps we were wrong about quasar distances, and they were much closer than we had thought so that the motions we saw were in fact below c. Or maybe the various sets of telescopes used for each observation had such different properties that it was misleading to try such a direct comparison of the data.

But as more observations of higher, and better–understood, quality accumulated, none of these ideas remained tenable. The motions we see, and in fact the very set of sources showing them, supported an interpretation springing directly from special relativity. It was obvious that something in the jets was moving outward, possibly near the speed of light. If we look down such a jet within a few degrees of its axis, light from a moving blob of material emitted at a certain time will arrive sooner after that from a previous time, compared to a stationary blob. This means that, within a critical angle depending on its velocity, we will see motions in a jet that can appear to go sideways faster than light. As a bonus, this idea explains why we should see such a bizarre phenomenon among the first few sources observed with VLBI. Relativistic approach velocities also produce *Doppler boosting*. The time transformation of special relativity compresses the object's time frame into ours, so we receive more photons per second than we would if it were stationary in our frame. The requirements for superluminal motion and Doppler boosting overlap; those most likely to show the superluminal illusion are also those whose brightness is most strongly boosted. Virtually every technique in astronomy was easiest to apply first to the brightest things in the

---

[13] A 1989 paper by Mark Reid and colleagues reported that the famous jet in M87 displayed not superluminal, but subluminal, motions. Inspection of the table of contents listed on the cover of that issue of the *Astrophysical Journal* suggests that the proofreaders might have been more accustomed to the behavioral sciences, since the paper is listed as "Subliminal Motion and Limb Brightening in the Nuclear Jet of M87". A subliminal motion might be one that happens too fast for conscious perception.

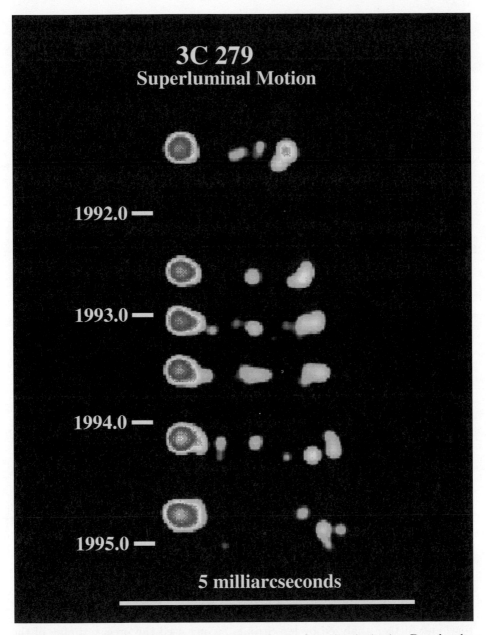

Fig. 2.13. The quasar 3C 279, showing superluminal motion in its jet. Pseudocolor intensity coding has been used to make subtle features more apparent. The prominent outer knots to the right are moving outward with a projected speed of 4c. At this quasar's distance of 5 billion light-years, the scale bar of 5 milliarcseconds corresponds to a length of 100 light-years. The bright knot to the right moves almost 20 light-years in 4.5 years, as viewed in our reference frame. These observations, at a radio wavelength of 1.3 cm, were carried out initially with an *ad hoc* network of radio telescope, and starting in 1994, with the Very Long Baseline Array of the National Radio Astronomy Observatory, a dedicated network of telescopes stretching from the Virgin Islands to Hawaii. (Images courtesy Ann Wehrle and Steve Unwin.) This figure also appears in color as Plate 1.

sky, and VLBI was no exception. With improvements in recording and correlating technology, it eventually became possible for VLBI to measure the weaker radio waves from sources not pointed at us (where any detected motion is often subluminal, since the line-of-sight effects do not swamp the transverse motion in our view). The final demonstration of this relativistic scheme had to await the discovery of smaller objects within our own Galaxy which show the same phenomena, and additional ones that take too long for us to see on the large scales of quasars. These are the microquasars (section 2.8).

2.8 Gamma-ray beams – small and large

Radiation from jets moving near the speed of light, beamed into narrow angles by this motion, has helped explain an astronomical mystery that took twenty years to uncover – gamma-ray bursts. They had the rare astronomical distinction of being discovered in a classified program, and having to wait years before migrating to common scientific knowledge. And uncovering even their true locations in space was to take two additional decades. This story begins in the coldest depths of the Cold War, in the aftermath of the Cuban Missile Crisis as American and Soviet leaders contemplated how close to nuclear war they had come. As a step toward at least mutual confidence, the Limited Test-Ban Treaty was signed among all the nuclear powers in 1963. They undertook to test no nuclear explosives in or above the atmosphere. Underground explosions would at least scatter no fallout, and would be virtually forced to have smaller yields than the mammoth fusion tests of the previous decade. Although the phrase "national technical means of verification" had yet to enter public discussion, the United States employed such means liberally in verification for the treaty. A series of *Vela* satellites was launched from 1963 to 1970, into such high orbits that they could keep watch not only on the territory of Earthly adversaries, but on much of the far side of the Moon as well. As a result of these high orbits (reaching 125,000 km from the Earth), most of the *Velas* remain in orbit (although some have been lost from tracking by now). Launched in pairs, each carried detectors for specific signatures of above-ground nuclear tests. In particular, they would look for the brilliant flashes of gamma rays, as well as visible light, that would be the unmistakable signs of nuclear explosions.

As the data streams from the *Vela* instruments were recorded, they occasionally included significant bursts of gamma rays from above rather than below. These data were collected by Ray Klebesadel at Los Alamos National Laboratory, and further analyzed by Ian Strong. Strong used the precise timing of the bursts as seen from various satellites (a fraction of a light-second apart) to show that, while their locations were very poorly determined, they did not originate from Earth or Sun. Klebesadel, Strong, and Roy Olson (also of Los Alamos) published their results in the (open) scientific literature in 1973. The four sentences of their abstract could also summarize most of what we learned in the following twenty years. These facts can be summarized very briefly. Bursts of high-energy radiation arrive randomly from unknown sources. Their durations range from shorter than 0.1 second up to minutes. Crude directional information eliminates obvious local sources.

With so little hard data available, theoretical explanations ran to wild ideas. The bursts could be nearby and comparatively modest in energy, or they might lie in distant galaxies and be the "largest explosions since the Big Bang". They could come from colliding neutron stars, or comets impacting the surface of neutron stars. Neutron stars were popular suspects because material falling freely onto their surfaces will impact at roughly half the speed of light, liberating enormous amounts of energy (and, unlike black holes, they have a surface to hit, so all that energy is released quickly and where we might see it).

Increasingly intricate (some might say "desperate") technical measures were employed to learn more about the locations and, above all, distances of the burst sources. The fundamental instrumental problem was that detection of gamma rays does not lend itself to most of the approaches that let us produce images with lower-frequency kinds of radiation. Gamma rays cannot be coaxed into reflecting from the surface of a mirror even at the very shallow angles that allow focusing of X-ray telescopes. Estimating the direction of a gamma ray basically amounts to letting it knock an atom loose in a detector and tracing the recoil backwards. As a result, the typical accuracy of burst positions did not improve beyond a degree or two until we could coordinate observations at other wavelengths rapidly enough to find associated emission at some wavelength that we can focus sharply. In the meantime, the delay between the burst times as seen from widely separated spacecraft was the best we could do. The Interplanetary Network incorporated a changing cast of probes in Earth orbit or en route to Mars or Venus, with small gamma-ray detectors that would pick up at least the brightest bursts. This provided one avenue for Soviet–US cooperation in space science, as data had to be shared for the idea to work – and the faster, the better. Eventually there were even exchanges in equipment; at least one Soviet *Konus* detector flew on a western spacecraft, after its counterparts had made numerous trips to Venus as well as more local regions of space. Triangulations from the Interplanetary Network yield crossing annular regions on the sky, whose intersection defines the position box of the source. Over and over, nothing special appeared in these intersections. That is, until the burst which reached the Solar System on March 5, 1979.

The March 1979 burst was unusually bright, being well detected by no less than fourteen spacecraft (including three of the *Velas* that had brought our attention to the whole phenomenon). The long timing baselines to a detector on the *Pioneer Venus Orbiter*, the *Venera 10* and *11* probes travelling to Venus, and *Helios 2*, in solar orbit, were crucial to pinning down its origin. And for once, there was something obvious in the position box – the remnant of an ancient supernova explosion in the Large Magellanic Cloud, about 160,000 light-years away. The best positional refinements fell entirely within the remnant catalogued as N49 (Fig. 2.14). Supernova remnants often host neutron stars, which are good places to release lots of energy; *ergo* this burst made the link between bursts and neutron stars. Or did it? A generation later, it appears that the March 1979 burst was a high-energy red herring, which may have set our understanding

Fig. 2.14. The supernova remnant N49 in the Large Magellanic Cloud, as imaged by *Hubble*'s Wide Field Planetary Camera 2. The box marks the region known to contain the source of the March 1979 gamma-ray burst – an identification which put astronomers off the track of more distant bursts for over a decade. Pulsars are known to acquire such rapid motions that it was quite plausible for one created in this supernova explosion to appear at the edge of the resulting nebula. The source was eventually understood as one of the so-called soft gamma repeaters, distinct in spectrum and repeating nature from the high-redshift bursts which dominate the gamma-ray sky. (Background image: NASA and the Hubble Heritage team, Y.-H. Chu, S. Kulkarni, and R. Rothschild.)

back by a decade. We know today that it fits in a different category: the soft gamma repeaters, which are indeed associated with highly magnetic neutron stars, much more common and less energetic than the majority of the bursts astronomers had been chasing. But this distinction was not at all apparent from what we knew in 1979.

The next step in at least the volume of burst data was expected from a large and very capable satellite. NASA's *Compton Gamma-Ray Observatory* (CGRO) was deployed from the Space Shuttle *Atlantis* in April 1991 (Fig. 2.15), joining

Hubble to become the second of NASA's four Great Observatories.[14] *Compton* needed help from astronauts Jerry Ross and Jay Apt, who freed a balky antenna in a rare unscheduled spacewalk before CGRO was released into independent orbit, functioning until fears of uncontrolled re-entry prompted its commanded plunge into the Pacific in mid-2000. At the time of *Compton's* launch, most people in the field expected it to confirm the association of gamma-ray bursts with neutron stars, and most of those to be in our own galaxy. While the individual position errors would be too large to verify identification with, say, known pulsars, the large number of bursts measured by the eight-sided BATSE system[15] could well show a concentration of bursts toward the Galactic Center or even the plane of its spiral arms – structure that would have evaded earlier instruments simply through sparse statistical sampling.

Compton's BATSE data showed nothing of the kind. It quickly became clear to the instrument team members that the expected results for neutron-star models did not match the incoming data. BATSE Principal Investigator Jerry Fishman notes that within three months of operation they could see patterns in the sky distribution and in the brightness of bursts that had them "both excited and nervous at the same time, since we knew the implications... We wanted to make damn sure in Huntsville that our analysis software was working properly and giving the correct locations." After measuring about one hundred BATSE bursts, the experimenters were confident enough to announce a distinct *lack* of any structure in their distribution across the sky – a result that only strengthened as this uniform catalog grew to contain 2,704 bursts (Fig. 2.16). In contrast, there was a highly noteworthy pattern to the statistics of bursts with intensity. Such statistics ("source counts") have a long and powerful history in new fields of astronomy. Objects of a particular luminosity uniformly scattered through three-dimensional space will show a characteristic increase in the number we can detect as we look a given factor fainter. Departures from this factor may be evidence of observational incompleteness, a nonuniform distribution in space, or, at large enough distances, a lack of objects or cosmic evolution in the average properties of the sources. Gamma-ray bursts showed a deficit of faint sources relative to bright ones. Either the bursts were in fact clustered closely around the Earth, which seemed to violate most of what we have learned about the Universe since Copernicus if not veer into downright paranoia about our occupying some special part of the cosmos, or most gamma-ray bursts are so distant that we are looking across most of cosmic history and watching their sources evolve, or simply running out of sources. The Earth-centered idea seemed to require such careful contrivance that ideas for very distant bursts, neglected for years by all but a few faithful scientists such as Princeton's Bohdan Paczynski, came

---

[14] The other two would be the *Chandra X-ray Observatory*, deployed from the Shuttle *Columbia* in 1999, and the infrared-sensitive *Spitzer Space Telescope*, launched in 2003 by a Delta II rocket. NASA administrator Sean O'Keefe called the quartet a Mount Rushmore of observatories.

[15] CGRO's Burst and Transient Source Experiment, with one detector on each corner of the spacecraft so that the entire visible sky was constantly in view to multiple detectors.

Fig. 2.15. The *Compton Gamma-Ray Observatory*, viewed from the overhead window of the Space Shuttle *Atlantis* shortly before its release into orbit on 11 April 1991. Four of the BATSE burst detectors are seen at corners of the structure. (NASA.)

back into vogue. By the mid-1990s, we knew more or less where the bursts came from. But what happened to produce these titanic outbursts, and how did they connect to the processes we see in other ways?

This answer was to require another new technological capability, as well as fast and hard work by teams of astronomers rushing to telescopes all over the Earth. The new technology was embodied in the Italian–Dutch satellite *BeppoSAX* (named in honor of the late cosmic-ray physicist Giuseppe "Beppo" Occhialini). *BeppoSAX* combined gamma-ray detectors with an imaging X-ray telescope, which could see the low-energy part of the burst emission. When a bright gamma-ray burst happened within its field of view, the instrument would show immediately where the associated X-rays were, to a level of precision fine enough to at least point a large ground-based telescope and know that it was somewhere in its field of view. This finally made it practical to look for some dim and fading optical counterpart or afterglow, and if found quickly enough, measure its visible-light spectrum. Such an observation would probably reveal the burst's distance and much about its properties. This capability was seized on, with several teams of astronomers ready to chase down potential counterparts. Within ten months of the satellite's launch, a fading optical counterpart had

2704 Gamma–Ray Bursts in Galactic Coordinates

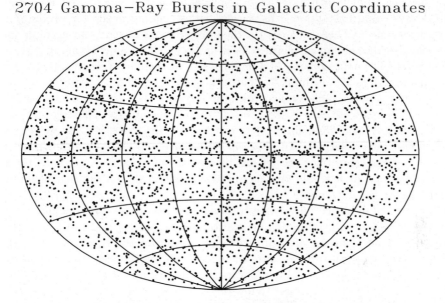

Fig. 2.16. Sky distribution of the 2,704 gamma-ray bursts catalogued by BATSE during the entire *Compton* mission, shown in galactic coordinates so that objects in the disk of our Galaxy would be concentrated toward the equator. No statistically significant pattern is seen. Individual bursts have a location error typically 2–3°, which may be compared with the grid lines spaced at 30° intervals. These data are available from http://gammaray.msfc.nasa.gov/batse/grb/catalog/current/.

been identified, and just over a year after launch, the first afterglow redshift was measured. The burst of May 8, 1997, had an optical counterpart which was observed by Shri Kulkarni and colleagues using the 10-m Keck telescope. It showed absorption lines at the phenomenal redshift of $z = 0.835$, so the source must be at least that distant. The counts had given the right answer – gamma-ray bursts come from phenomenally large distances, and must be correspondingly powerful. As of early 2005, redshifts have been measured for nearly fifty bursts, reaching as far as $z = 4.5$ (which also ranked its surrounding galaxy among the most distant measured). When we can see the galaxies around them, often after the glow fades, they are usually blue galaxies with active star formation. It has been our good fortune that many of these bursts have associated afterglows that fade much less rapidly than do the gamma rays, perhaps being produced as rapidly expanding material from the burst plows into surrounding gas.

Theorists were scrambling to fit these pieces into a coherent physical picture. The energy radiated in gamma rays alone would have to be far beyond that of any process we knew of – assuming, that is, that the gamma rays went out in all directions. The energy requirements would become easier to understand if they all went out only within some small angle from the source – which is to say, if the gamma rays were beamed rather than emitted in all directions. Beaming would solve another problem if it came from relativistic motion, since what we see as gamma rays could have started as X-rays, whose energy needs for production are easier to meet. There was even a theoretical scenario ready to account for this – the so-called collapsar picture. As it has been elaborated by many people, this scheme follows the collapse of a massive star at the end of its life. Most such collapses would yield "ordinary" supernova explosions. Some, however, would form a black hole or massive neutron star temporarily surrounded by a disk of matter about to fall into it, *deep inside the collapsing star*. Lasting only for seconds, such a disk could funnel jets outward in the same way as we see in quasars and their relatives. An observer situated along the axis of such a jet would see gamma rays once the jet has pierced the star's surface; outside these favored locations, little would be seen until the jets had slowed, so their beaming would become less extreme.

Two lines of evidence have come to favor this beaming picture. One comes from the collection of brightness information on the optical and radio emission sometimes found from the afterglows of gamma-ray bursts, the same emission that allowed astronomers to measure their redshifts (and indeed find them with precision). Many of them not only dim rapidly, but pick up the pace of this fading within a few days of the outburst (a few days, that is, in our frame of reference). This break in the intensity behavior could come about as a relativistic jet slows down on encountering surrounding material. Early on, when we see light from a relativistically expanding object, we see the same thing whether it is a narrow jet or a spherical fireball. We will see only the small part that is moving most directly toward us and Doppler boosted to high brightness – so even if it is a fireball, from our point of view it might as well be a narrow jet. As it slows down, each part of the object emits light beamed into progressively wider areas. If it is indeed a jet, there will come a time when, as we see more and more of the entire jet's size while its beaming broadens, there will be no more of the jet to see because we can now see all of it. After that, as its beaming continues to relax, it fades more rapidly, as the jet's light is spread over a wider and wider region of space. In contrast, for a round fireball, there is no such distinct change in beaming behavior, only a smooth light falloff. Such events occur in many well-observed bursts. Because gamma-ray bursts do not announce their intentions, being sure the right measurements are obtained can be an exercise in rapid-fire negotiations. I recall getting a phone call late one night at an Arizona observatory, from a colleague who had just gotten off a plane in New York (where it was already about 2 a.m.), and checked his email to find notice of a bright burst. He was now frantically lining up optical and infrared measurements of a poorly-known location before that part of the sky set as seen from the western U.S. Fortunately his efforts paid off, yielding a timely set of optical and infrared data as the object faded away over only a few days. The conclusion may be gleaned from the title of our resulting journal paper, "Discovery and Monitoring of the Optical Afterglow of the Energetic GRB 991216: Joining the Jet Set".

The afterglow data suggest that some kind of jets are involved – a welcome enough conclusion since it reduces the energy emitted by the bursts by perhaps a thousand times over what we would think if they sent the gamma rays equally in all directions. More specific evidence which implicates the deaths of (some) massive stars in producing the bursts comes from the association of some bursts with supernova explosions. The most direct link came from the burst identified as 19980425, at the remarkably small redshift of $z = 0.0085$ (a mere 115 million light-years away, the most local example yet found). This burst afterglow was coincident with the supernova catalogued as SN 1998bw in a spiral galaxy. The supernova showed signs of being unusual in several ways. The explosion itself produced stronger radio emission than any yet observed, and its optical spectrum was of the relatively rare Type Ic. These are distinct in showing even faster ejection of the debris than usual – as much as a tenth the speed of light – and the chemical makeup of a star whose outer layers had been blown away by a powerful stellar wind well before the explosion. Detailed calculations suggested that such a supernova could result from a massive star whose core collapsed directly to a black hole during the initial collapse that sets off the surrounding explosion, without forming a hot neutron star. This fits with the collapsar picture, in which the jets are funneled by a temporary disk of stellar material around this massive core. The burst of 1998 was not unique in this regard; close examination of the sites of a few later bursts as the glow of the initial flash fades has reveled the spectral lines of type Ic supernovae. Additional evidence that gamma-ray bursts come from supernova sites has come from the X-ray spectra measured by the *Chandra* observatory; in one well-observed instance, spectral features indicated a high abundance of just those heavy elements formed during a supernova explosion. The association of these bursts with some kinds of stellar explosions seems secure, although much remains to be known. What kinds of stars are the progenitors? How are the jets accelerated so rapidly? How many supernovae look like gamma-ray bursts to someone, somewhere in the cosmos?

The dynamic gamma-ray sky shows us additional kinds of beams from the depths of space. In addition to the key role that *Compton* data played in pointing to a distant origin for gamma-ray bursts, the observatory's main instruments surveyed the entire gamma-ray sky with unprecedented sensitivity, cataloguing hundreds of gamma-ray sources. As so often happens, major effort has gone into identifying these gamma-ray sources with objects that are familiar from other windows in the electromagnetic spectrum – efforts which are complicated by the modest position accuracy of the gamma-ray observations. As a result, many of these sources remain unidentified. Among those that have been firmly linked to known objects, many turn out to be a specific kind of quasar – *blazars*. Blazars are distinguished by several features which suggest that they contain relativistic jets pointed almost toward us – rapid variability, a continuous spectrum which is unusually strong compared to emission lines from the surrounding gas, and radio structure dominated by a very strong pointlike core rather than a large double source.

Quasars appearing in the *Compton* catalogs are overwhelmingly of the blazar clan. In hindsight, this makes sense – the enormous boosting in frequency and arrival rate of the photons from an approaching jet, which emits plenty of X-rays in its own frame, would make such objects bright gamma-ray sources when seen in our reference frame. Looking down the jet adds another kind of energy boost, since this is the direction where scattering between X-rays and the relativistically moving particles in the jet can add further energy to the photons. In some gamma-ray blazars, half the energy we see arrives in gamma rays, in a high-energy echo of the whole normal energy distribution of the object (Fig. 2.17). In fact some of these gamma rays have energies too high to be detected by our satellites.

After all the remarkable discoveries we have made with the opening up of our views in more and more parts of the spectrum, most of which are in wavelengths that do not penetrate our atmosphere, it is strange to mention high-energy radiation that our satellites cannot see. But what sets the limit here is not energy itself, but the rarity of the most energetic photons. As we look at cosmic sources by detecting gamma rays of ever-higher energy, each source sends us fewer photons (even though each photon packs more energy and contributes more to the overall intensity). A detector of fixed size will thus see fewer and fewer, until finally none can be detected at all in even the longest possible exposures. Even a chunk of metal the size of a small car is not big enough to grab these rare events. But our atmosphere, the bane of astronomers, comes to the rescue by acting as an enormous extension of our detectors. Just as energetic particles in cosmic rays can interact with atoms in the upper atmosphere and trigger showers of particles produced from their energy (Chapter 3), so can the most energetic photons. A gamma ray of high enough energy, on reaching the outer levels of the atmosphere, can discharge its energy in the creation of pairs of particles and antiparticles, most of which will be rushing downward near the speed of light (to preserve the cosmic bookkeeping by conserving both momentum and mass+energy). *Very* near the speed of light, in fact. Many of these secondary particles will be so close to c that they exceed the speed of light in air, which slows visible light by about 0.03% (near sea level and even less at high altitudes). Recalling that special relativity tells us that no object can exceed the speed of light *in vacuum*, we see that no law of physics is violated by an object that squeezes into this tiny velocity range between the speed of light in air and in vacuum. But even so, exciting things will happen. The particle sets up a kind of shock wave in the air, losing energy and generating a cone of blue light – Cerenkov radiation. The faster the particle, the more tightly the cone of radiation can be seen only looking back along its track. For particles with enough energy, the result is that we see a brief flash of light when we look back in the direction of the original gamma ray – a flash that lasts only a billionth of a second and may be seen from a region a few hundred meters in size. This we can look for from the ground, by using "light buckets" – collecting mirrors of large size but requiring very little in the way of optical quality (Fig. 2.18). In fact, some projects use the towers and arrays of flat mirrors designed as large solar ovens at night, substituting photomultipliers for the daytime material to be melted. This technique gives us a window into the very highest energies of the electromagnetic spectrum, in which the brightest gamma-ray blazars can be seen even at energies where we are decades from building large enough satellite detectors. That makes the observations triply relativistic. The blazar jets are bright gamma-ray sources because they are coming toward us so fast, and the photons are produced in a

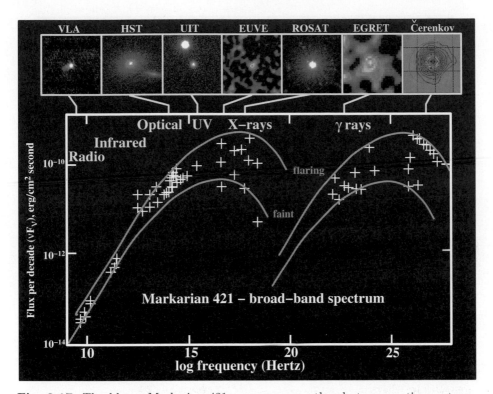

Fig. 2.17. The blazar Markarian 421 as seen across the electromagnetic spectrum. This montage shows the relative energy received from this blazar in various spectral regions, both when it is quiet and when it is in a bright flare such as may accompany the appearance of new jet material. The inset images show its appearance to various instruments used for these observations. The distinct gamma-ray peak is evidence that its lower-energy radiation is being scattered and Doppler-boosted in a jet directed nearly toward us. (Data provided by T. Weekes, the NRAO FIRST radio survey, and the NASA archives from the *Hubble, Compton, ROSAT, UIT,* and *Extreme Ultraviolet Explorer* missions.) This figure appears in color as Plate 2.

process relying on the relativistic energies of their internal particles, measured using a relativistic interaction in our atmosphere.

The gamma-ray blazars are also systematically those that show superluminal motion in their jets. The sky furnishes us with a surprising range of objects in which we see apparent motions faster than light, because of the relativistic compression of time coming into our reference frame. Alabama colleague Gene Byrd has pointed out that this compression of time, as some object comes toward us so fast that it almost catches up with its own light, becomes clearer with a more cinematic formulation. At a close approach between Earth and Mars, their separation can be as small as 55 million kilometers, or 183 light-seconds. Imagine some vast alien spacecraft leaving Mars for our vicinity – one so huge that the

Fig. 2.18. The setting Sun is reflected in the three mirror arrays of a gamma-ray Cerenkov system, erected by a group from the University of Durham and used in the 1980s at the Observatorio del Roque de los Muchachos on the Spanish island of La Palma.

flash of its launch would appear in our telescope, and able to travel at 90% of the speed of light.[16] The spacecraft would take $183/0.9=203$ seconds to flash past us, but from our naive observational point of view, we would see it arrive only 20 seconds after we saw it launch, so, from that point of view, we would measure its speed not at 90% of the speed of light, but as over $9c$. This is the basic reason that relativistic jets can appear to move faster than light, as long as they are moving nearly enough toward our location.

2.9 Microquasars have minijets

For years, the presence of relativistic motions in jets streaming outward from black holes was a purely theoretical notion, if one that explained several otherwise puzzling facts. In 1978, a remarkable confluence of results from X-ray, radio, and optical astronomy finally presented astronomers with this process happening before their eyes, leaving no doubt that such jets can occur. The scene was not in some enormous and active galaxy a billion light-years away, but in the remnant of a binary-star system only a few thousand light-years from home. The trail of its discovery is marked by false starts and missed opportunities, and could equally have begun from a routine survey of the spectroscopic properties

---

[16] It is one of the pleasures of thought experiments that such details as acceleration and deceleration can be completely ignored unless they are relevant to the argument.

of stars on photographs, from a radio-frequency study of the suspected remnants of supernova explosions, or from X-ray surveys of interesting objects identified with stars. The path that led most directly to it started with attempts to find the leftovers of supernova explosions – the tiny, collapsed cores that might be either neutron stars or black holes (see Chapter 7). The discovery of pulsars in 1967 had shown that some supernovae leave behind rapidly spinning neutron stars, but the possibility remained (and remains) that some ways to blow up a star leave different kinds of object. Checking on this possibility, Jim Caswell and David Clark looked at a suspected supernova remnant known as W50, using the Molonglo radio telescope in Australia – a cleverly designed, gigantic cross-shaped antenna 1.6 kilometers in extent, that can be electronically steered toward different parts of the sky without physically moving the antenna. Their map showed an arc marking one edge of the huge bubble blown in the interstellar gas by an ancient explosion, and a bright patch near the center of the arc. This patch attracted interest, since it might represent a new kind of leftover from a supernova explosion – a remnant producing steady radio emission, rather than the brief and repeating flashes of pulsars.

Two years later, this same "patch" attracted attention first as an X-ray source, seen by the UK's *Ariel V* satellite, and also when its radio emission was seen to flare erratically – a sign that the radio emission had to come from a small source rather than some huge diffuse cloud of matter. These continued the strong role of British astronomers in solving this particular astronomical mystery – the X-ray studies involved Fred Seward from the US working with the strong X-ray group at Leicester University, while the radio studies were led by John Shakeshaft at Cambridge.

By the time of an international conference in May 1978, held in the Sicilian town of Erice, the question of just what can be left behind when a star explodes was coming into better focus. There were hints from radio telescopes that too many supernova remnants had small radio sources (of the nonflashing variety, distinct from pulsars) at their centers to be attributed to chance superpositions of unrelated background objects. Various teams picked different examples to follow up with other techniques. Ironically, most of these turned out to be unrelated background objects, so that some teams gave up the hunt in discouragement too early. Clark and Murdin picked the supernova remnant W50 for further study.[17] This source, in the middle of an elliptical ring revealed by new radio observations, was apparently identical to the X-ray source known from its position as A1909+04. The position of both the radio and X-ray source was still rather poorly known by the standards of optical astronomy, so that it remained to be determined which, if any, of many stars close to that location might be identical with the source of the energetic radiation seen in other windows.

Within weeks, Murdin and Clark were able to practice their search strategy, in a session using the 4-meter Anglo-Australian telescope in New South Wales.

---

[17] "Supernova remnant" here means the expanding, hot bubble of debris from the supernova explosion, rather than the compact remnant object at the center – neutron star or perhaps black hole – often remaining from such events.

Among the generation of 4-meter telescopes brought into operation in the 1970s, the AAT distinguished itself by its accuracy of operation and advanced detectors – features which could help make up for its site being lower and less stable than some of its competitors in Chile and Hawaii. They had originally planned to observe extended nebulous supernova remnants, but a failure in one of those advanced detectors forced a switch to a different spectrographic system – well suited to examining individual stars rather than extensive regions of the sky. On the evening of June 28, they took a close look at the environment of A1909+04, starting with the latest radio coordinates provided by Shakeshaft. Suspecting that any collapsed remnant of a supernova would be tiny and faint, they started with various dim stars near that position, putting off until the end a surprisingly bright star (Fig. 2.19) within the margin of error for the radio source's location. When they finally looked at this "obvious" candidate, a peculiar spectrum with strong emission peaks of glowing gas appeared on the monitor. Murdin must have felt as much triumph as astronomers ever do in the dome as he called out, "Bloody hell! We've got the bastard!"

The spectrum was, at first glance, much like another X-ray source associated with a supernova remnant – Circinus X-1. They could see emission peaks at wavelengths roughly corresponding to hydrogen, helium, and a few other elements. Few stars are surrounded by the kind of tenuous, excited gas needed to produce such spectral features, so any star showing these spectral peaks is automatically unusual and interesting. But an accident of timing kept Clark and Murdin from appreciating just how unusual and interesting their discovery was. At least the identity of the X-ray source was settled; it was a star previously catalogued by Bruce Stephenson and Nicholas Sanduleak as showing hydrogen emission, number 433 in their list. It was thus to reach fame under the name SS 433.

On returning to England, Clark met Bruce Margon, near the end of a visit from his home base at UCLA. Margon had a long-term interest in X-ray stars, pursued with drive and ambition. SS 433 piqued his interest; before an observing run at Lick Observatory's 3-meter telescope in September of 1978, he asked Clark for more details and an identification chart. Margon's observations at the end of September were, at once, to deepen the mystery of SS 433 and provide the keys to its solution.

The University of California has long been an astronomical powerhouse, with major astronomy departments located at four campuses and its own large observatory, Lick Observatory near San Jose. In the late 1970s, when large telescopes were even scarcer than today and the 3-meter reflector (Fig. 2.20) ranked as the sixth largest telescope on the planet, UC faculty thus enjoyed unique access to large amounts of telescope time on a regular basis, at a location blessed with the favorable climate of the coastal California mountains and frequently excellent image quality. This location does have a drawback. In the century since the observatory was established from the bequest of James Lick, this same climate had attracted millions of people, so that Lick Observatory today looks down on the lights of the entire San Francisco Bay area (Fig. 2.21). This handicap has forced Lick astronomers to push a series of technical advances to keep the observatory

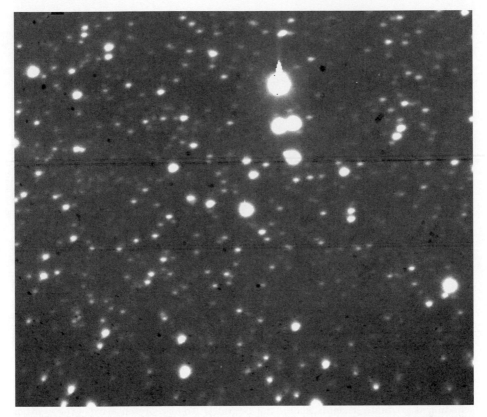

Fig. 2.19. SS 433 is the bright star at the center of this image, which shows the abundance of faint stars also allowed by early X-ray positional uncertainties. This entire image lies well within the supernova remnant W50. Data from the Lowell Observatory Hall 1.1-meter telescope.

competitive and useful (especially since Caltech operates the larger 5-meter Hale telescope at Palomar just down the coast, and in a darker environment). Among these developments were the Image-Dissector Scanner (IDS), developed by Joe Wampler and Lloyd Robinson to provide a system for digitally subtracting the light of the night sky, and showing the results in real time. A later version of the IDS had been built for the Anglo-Australian Telescope, and used for the first detection of SS 433's oddities.

On observing the spectrum of SS 433 with the IDS system at Lick, Margon immediately saw the strong emission lines that had convinced Murdin and Clark that this was something special. But, looking in more detail, he also saw additional emission peaks, at wavelengths not correponding to any obvious features in stars. This was something new. There are patterns of spectral lines that go together, and only a small number of lines that are ever found to be strong. Yet SS 433 showed strong spectral features at completely random wavelengths - something Margon described as a "disquieting state of affairs", vaguely as if

Fig. 2.20. The 3-meter (120-inch) Shane telescope at Lick Observatory, used by Bruce Margon and colleagues to trace the unique changing Doppler shifts of SS 433. The exterior view, taken on the one night that two of us had our names on the observing schedule as graduate students, should not be construed as representative of the weather on Mount Hamilton.

Fig. 2.21. The San Francisco Bay area as seen from Lick Observatory beyond moonlit hills. The light pollution produced by these vast cities forced Lick astronomers to pioneer electronic techniques for subtracting interfering skylight from the spectra of faint objects.

you were to flip through the New York phone directory and find random pages printed in Cyrillic or Hebrew letters – or some alphabet you had never seen at all.

SS 433 is bright enough that useful spectra could be obtained with much less powerful means than the 3-meter telescope; in fact, it is within reach of today's well-equipped amateur spectroscopists. An earlier version of the IDS was attached to a 60-cm telescope at Lick, with which resident observer Rem Stone followed SS 433 for four nights later in October 1978. The lines moved! This was unprecedented in astronomical history. Spectral features were wandering back and forth in wavelength by enormous shifts (Fig. 2.22). Margon's team tried all sorts of ideas to avoid the most obvious explanation (championed by Stone) that these were Doppler shifts, and that we were seeing material not only moving at a good fraction of the speed of light, but changing either in that speed or its direction from one night to the next. What pinned down the Doppler explanation was the fact that the spectra showed red- and blue-shifted versions of several spectral lines, and that the shifts for all the lines matched exactly if they represented rapid motions. It may be just as well that I cannot recall which of the Lick staff astronomers saw an early report by Margon's group on a bulletin board and was heard to mutter, "Margon has broken the spectrograph!" The oddities of SS 433 attracted media attention – perhaps most memorably when the character Father Guido Sarducci, in a *Saturday Night Live* TV comedy sketch, spoke of the star which was "coming and going at the same time".

The study of SS 433 cried out for more data, and soon astronomers worldwide were contributing spectra. Eventually there was evidence of a pattern in the variation. Plotted against a 164-day period of variation, the Doppler shifts showed a repeating "lip" pattern, with two sets of spectral features changing in opposite directions, but in lockstep. Twice in each cycle, they crossed, with the "red" features temporarily heading to the blue side and vice versa.

Margon announced this bizarre behavior at the next Texas Symposium on relativistic astrophysics (in Munich).[18] Theorists swarmed, having strange and wonderful new things to explain. Within a few weeks, the correct explanation was found in papers by Martin Rees and Andrew Fabian from Cambridge (UK) and, with some additional wrinkles, by Mordehai Milgrom in Israel. Both suggested the same basic features. A black hole, located in a binary system with a more ordinary star, is accreting material, which gives off the strong X-rays as it is heated by the high densities near the black hole. Some of this material is sprayed off in a pair of narrow jets, possibly by magnetic fields, and shot away as we see in quasar cores. Unlike the case in quasar jets, these jets contain gas cool enough (at a few tens of thousands of degrees) to produce well-known emission lines and give distinct peaks so that Doppler shifts can be measured. In SS 433,

---

[18] It was a sign of the intellectual maturity of relativistic astrophysics when a series of conferences was organized specifically to exchange ideas at the interface of astrophysics and relativity. The first, in 1963, was organized by a group of physicists from the University of Texas in Dallas, and the name 'Texas Symposium' stuck despite most of the subsequent meetings being held in such locales as Chicago, Jerusalem, and Florence.

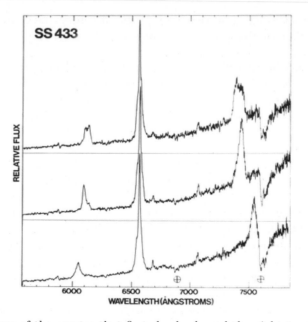

Fig. 2.22. Three of the spectra that first clearly showed the night-to-night motion of emission lines in SS 433. These are from three of four consecutive nights in October 1978, as observed by R.P.S Stone using the original IDS spectrum scanner on the 61-cm telescope at Lick Observatory. The highest peak is Hα emission from hydrogen, flanked by its blue-shifted (left) and red-shifted components. Each moves to a more extreme Doppler shift in each spectrum from top to bottom. Circled cross "Earth" symbols indicate wavelengths absorbed by oxygen in our atmosphere. (Courtesy of Bruce Margon.)

the Doppler pattern fixes the jet velocity at 0.26 times the speed of light. A misalignment between the accreting disk of gas and the orbit of the companion star (which is to say, between the black hole's spin and companion star's orbit), drives a precession so that the jets sweep out cones in the sky every 164 days (accounting for the changes in Doppler shift we see as the jets change direction). Finally, Milgrom pointed out that even the "ordinary" and "unshifted" features in the spectrum of SS 433 exhibited a long-predicted effect of relativity which astronomers had never seen in such an obvious way. These features actually show a redshift, so large that if interpreted as an ordinary Doppler shift from the motion of SS 433, this star would be moving at a healthy 12,000 km/s out of the Milky Way (whose escape velocity near SS 433 is closer to 450 km/s). However, since we see that material in the jets is moving so fast, other effects that Einstein wrote about become important. In particular, we must see time pass more slowly in the jets' frame of reference, so that all their radiation comes to us stretched in wavelength by a ratio (the familiar Lorentz factor), on top of the shifts due to their motion toward us or away from us. This means that the "crossing points" when both jets are sideways to us, and the wavelengths of any additional gas that has the same velocity, will appear shifted to the red by this factor – as we actually see.

These schemes for precessing jets could take the existing data for shifts of the spectral lines in SS 433 and predict what the wavelengths should be into the future, as the cycle of jet precession continued, or indeed run the clock back to earlier dates. This exercise showed how Murdin and Clark could have missed such an odd pattern of spectral lines. When they first recognized the star using the Anglo-Australian Telescope, the redshifted jet produced its strongest feature beyond the redward edge of the piece of spectrum they were observing, and other spectral features were, by chance, almost blended with "ordinary" ones. It is one thing to be outsmarted by a competitor, but quite another to be outsmarted by the Universe!

Global efforts confirmed this double-jet picture and filled in some of its details, while introducing us to further puzzles. The jets are not quite smooth, consisting of blobs that we can recognize in the visible-light spectrum because successive ones have slightly different Doppler shifts. This means that if we look at the spectrum closely from night to night, the "moving" lines are made up of pieces that are fading at one wavelength while new pieces appear at a slightly different wavelength. Recent studies have suggested that these "bullets" start out with a wide range of Doppler shifts, but always centered on the average value given by the 164-day cycle. This fact might be telling us something fundamental about how the jets are accelerated to such high speeds in such narrow cones, if we only knew how to read the clues. Similar blobs appear to radio telescopes, which can be set up with high enough resolution to actually see the jets and watch the blobs move outward from the central object. By 1980, astronomers had used several continent-spanning networks of radio telescopes to peek deep inside SS 433, finding twin jets of emitting knots flying away from the central source at the same $0.26c$ inferred from the optical spectra. In fact, the receding jet seemed to be moving a little slower, a light-travel effect since its radiation constantly reaches us from farther and farther away.

Other observers concentrated on the enormous gaseous shell W50 surrounding SS 433, seeking clues to what kind of supernova explosion might have left such exotic debris. In fact, at the outset it was not obvious whether this was the aftermath of a supernova explosion, until new images from wide-field Schmidt telescopes in Australia and California revealed networks of ionized gas that could be examined spectroscopically. The radio appearance of W50, in particular, showed something seen in no other supernova remnant – two "ears" punching out of the usual round shell (Fig. 2.23). These are lined up with the direction of SS 433's jets as seen by the global radio arrays – signs that the jets continue outward with significant impact as far as a hundred light-years from SS 433 itself.

These jets contain an intriguing mix of hot and cold matter. The very fact that we see strong emission lines from hydrogen in the visible spectrum means that they contain gas at a few tens of thousands of degrees close to SS 433. This material vanishes rather quickly; *Hubble* images do not show any hints of the jets, so their visible glow fades within light-hours. In contrast, orbiting X-ray telescopes show spotty trails (Fig. 2.24), in this case energized as the jets

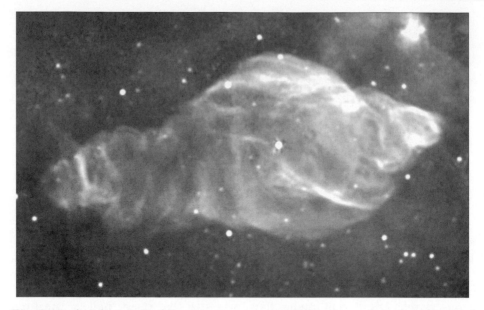

Fig. 2.23. A radio image of the supernova remnant W50, observed at a wavelength of 20 cm with the Very Large Array. The bright source of SS 433 itself is at the center, surrounded by the "bubble" blown by the ancient supernova explosion and the "ears" to east and west pushed outward by the jets from the central object. The elongated shell spans 1°.5 in our sky, making it a challenge for many modern radio telescopes to observe the entire structure in detail. This is a composite of multiple fields. (Courtesy Gloria Dubner, Mark Holdaway, and NRAO.)

strike the rarefied surrounding interstellar gas, seen extending outward for tens of light-years as they push outward on the supernova shell to produce the "ears" that distinguish its radio structure.

SS 433 provided proof of (at least mildly) relativistic jets from an object whose distance is not in serious question, as quasars were in the minds of a few astronomers. In fact, the relativistic light-travel effects in its jets provide a very accurate distance from strictly geometrical arguments, which is something to be highly prized. Radio and X-ray surveys suggested that there was nothing else in the Milky Way or even the Andromeda galaxy quite like SS 433. Still, our Galaxy had further tricks to show us, continuing to hide even more energetic objects that could act out scaled-down versions of all the tricks available to quasars and radio galaxies. The discovery of these *microquasars* once again called for a marriage of techniques using very different parts of the electromagnetic spectrum.

The spectral bands best able to penetrate the clouds of interstellar dust which block other views of most of our Galaxy are the radio, all but the lowest-energy X-rays, and gamma rays. Experience has shown that it is almost always interesting to examine counterparts of X-ray or radio sources at other wavelengths, and this is particularly so for the numerous sources in the plane of the Galaxy and

Fig. 2.24. The extensive X-ray jets of SS433, as seen by the *ROSAT* X-ray observatory. The eastern jet, approaching us as known from its motion and intensity, stretches across nearly a full degree of sky. Data from the *ROSAT* archive maintained by the Max-Planck-Institut für Extraterrestrische Physik.

near its core. They are mostly black holes or neutron stars being fed by binary companions, and thus laboratories for the most extreme phenomena (see Chapter 7). Indeed, some of these objects are small-scale analogs to the spectacular output of quasars, from gamma-ray emission to jets emerging near the speed of light. And because we know their distances reasonably well simply from their locations within the galaxy, they can furnish stronger demonstrations that we really are seeing relativistic optical illusions.

In hindsight, SS 433 might be allowed as a member of the class of microquasars, even though its jets travel at a mere quarter of the speed of light. But the definition actually arose from a more extreme instance. In 1992, long-time collaborators Luis Rodriguez and Felix Mirabel selected a group of newly located X-ray sources near the plane of the Milky Way for study at radio wavelengths. Among these was 1E1740.7-2942 (the E denotes, fittingly, that it was first discovered by the X-ray detectors on the *Einstein* observatory). But the key data in knowing exactly where to look came from yet higher energies than *Einstein's* telescope could focus.

Despite successes with focusing X-rays into images using mirrors aligned at very shallow angles to the radiation, localizing sources of higher-energy radiation had proven to be a daunting challenge. Mirrors do not work for high-energy (so-called "hard") X-rays or gamma rays. A variety of tricks have been used to precisely locate these sources of emission. If the same source is bright in soft X-rays, which can be focused by cleverly nested mirrors, an observation by one of the imaging X-ray observatories - *Einstein*, *ROSAT*, or *Chandra* – can locate it to arcsecond accuracy. If not, we must rely on use of the high-energy radiation itself. Some instruments use the techniques of nuclear physics, backtracking the recoil of particles or entire atoms to estimate the direction of the incoming photon. Others use a masking scheme, elegant in its simplicity but becoming progressively more subtle in its realization. A carefully shaped mask, composed of alternating segments which are transparent or opaque to gamma rays, is placed far from a detector array (2–3 meters is typical). A gamma-ray source at a particular location in the sky will cast a shadow of the mask, offset according to where the source is compared to the telescope's pointing direction. Use of a mask with openings of various sizes can provide unique patterns for each possible source location (unlike a simple checkerboard pattern, which gives the same observed pattern for many locations). Such a coded-mask device was used on the French SIGMA telescope carried as part of the payload of the USSR's *Granat* satellite (Fig. 2.25), operating in high Earth orbit throughout 1989-97. Given enough time (in pointing at the Galactic Center, SIGMA amassed 9 million seconds or about three and a half months) this arrangement produced an image of the gamma-ray sky with a resolution of $0°2$, and could localize the centers of bright sources to $0°01$.

This localization at higher energies, thus collecting the most energetic (and often highest-temperature) objects, and the ability to find those whose high-energy radiation was widely variable and hence from a small and active object, gave Rodriguez and Mirabel the final selection they needed in their hunt for exotic objects even among the X-ray sources already mapped by *Einstein*. When they turned the Very Large Array (VLA) in New Mexico on 1E1740.7-2942, the radio data revealed twin jets emerging from the core in opposite directions, like a junior version of radio galaxies and quasars. Its X-ray emission had the general spectral characteristics predicted for a hot disk of gas around a black hole. They reported this discovery in the journal *Nature*, and proposed the term "microquasar". The journal featured the word on its cover, and it entered the astronomical lexicon.

Fortunately, the class contains more than one member. Additional, and even more striking, examples soon appeared with observations of energetic galactic objects, particularly those from the *Granat* survey and data taken by the *Compton Gamma-Ray Observatory* (CGRO). By the end of 1992, Alberto Castro-Tirado had drawn attention to the *Granat* source GRS 1915+105, and teamed up with Rodriguez and Mirabel for VLA observations to probe its nature.[19] This object showed not simply a double radio source, but twin clouds measurably shooting away from the central source. Absorption by cold hydrogen, which is concentrated in our Galaxy's spiral arms, pinned its distance at about 40,000 light-years – well out on the other side of the Galaxy. Knowing its distance, they could tell how fast the clouds were leaving the object. The result, just as in many bright radio galaxies and quasars with jets, was an apparent velocity faster than light. This suggested, of course, the same explanation – motions close to the speed of light directed almost toward us, so that the clouds nearly catch up

---

[19] Rodriguez, from Mexico, and Mirabel, Uruguayan by birth, shared the 1996 Rossi Prize of the American Astronomical Society for their work on microquasars. After years sharing his time between France and Argentina, Mirabel was recently named the representative in Chile of the European Southern Observatory, with responsibility for both scientific and administrative matters of a large international organization.

with their own radiation. But in this case, we could confirm this with a clarity impossible for the more distant quasars and their kin. Both jets (or clouds), if moving outward at the same speed, suffer the same time dilation simply due to their motion, but their Doppler shifts will be enormously different. This means that the approaching one appears both dramatically brighter, and faster moving, than its receding counterpart, as we see its timescale speeded up by the relative motion. If the two sides of the source are in fact the same, the brightness and speed ratios together uniquely determine the viewing angle and ejection velocity. If the ejection velocity and angle are within certain ranges, the object will appear to us to be expanding faster than light. *For microquasars, we can therefore show that their jets must be very near the speed of light and pointed nearly in our direction.*

Fig. 2.25. The *Granat* satellite, photographed during final mating to its Proton booster at the Baikonur cosmodrome. The SIGMA coded-mask telescope fills the large central tube. Most of the satellite name in Cyrillic, *ГРАNAT*, can be read on the payload shroud. Image courtesy of CESR, Toulouse, France, provided by Laurent Bouchet.

With such strong hints of what to look for, astronomers worldwide rapidly filled out this picture. Within a year, a third microquasar, GRO J1655-40, had been reported by groups in New Mexico and Australia. This object showed very active superluminal jets, beautifully illustrating the asymmetry from Doppler shifts and Doppler boosting. By now more than twenty microquasars have been recognized in the Milky Way, each containing a compact object (that is, neutron star or black hole) and twin jets. All these jets are relativistic, but not all are superluminal. This is in line with the superluminal jets requiring special viewing locations – for most of them, we're not in the magic cones to see this grand illusion.

Microquasars teach us something else about relativistic jets, something we might have to wait millennia to see in quasars. There is a frequent pattern of the X-ray brightness fading and reappearing just before radio telescopes see new, bright blobs of jet material. This may show increases in the inward flow of material, damping the escaping X-rays, shortly before some of the material is flung off (magnetically?) to form the jets. Because the scale of the black holes and their surrounding disks is so much smaller, we last long enough to see this process in microquasars – not just once, but over and over.

In addition, we can sometimes see the "normal" stellar companions that feed the black holes or neutrons stars at the heart of microquasars. Our view may be confined to the infrared by intervening dust, but even there we can learn enough about the companions to estimate the masses involved (and tell which ones have to be black holes). One of these, XTE J1118+480, is found to orbit well out in the halo of our Galaxy. It may well have come from a globular star cluster, having been liberated from the cluster by a close encounter with another star. Such encounters can also leave behind a very tightly bound binary star, so the same event could have been responsible for both the existence of this system and its motion around the Galaxy.

2.10 The Crab does the wave

The motions that give rise to relativistic distortions need not involve actual motion of matter near the speed of light. A wave motion through such matter can do just as well, as seen in light echoes. An example was revealed when *Hubble* and *Chandra* took a long, coordinated look at the Crab Nebula.

The Crab Nebula is the expanding debris of a supernova explosion seen in AD 1054, visible in daylight for weeks. The Crab has played a leading role in so many aspects of astrophysics that it was often said (at least by astronomers studying this object) that there are two branches of the field – studies that dealt with the Crab, and all the rest. Here we see synchrotron radiation, extending to the highest energies we can measure, a central pulsar whose magnetic field drives a powerful wind outward through the nebula, instabilities condensing surrounding gas into fingers as the pulsar wind encounters it in an oil-and-vinegar combination, relativistic jets from the pulsar decelerating in ways we would never live long enough to see in radio galaxies... The Crab is a full-course meal with something for every palate. Recent observations with *Hubble* and *Chandra* have reworked our picture of this familiar object, and show us why it still appears so bright and dynamic after nearly a millennium.

The best ground-based pictures of the Crab have long shown a pattern of wisps near the central pulsar, whose appearance changes with time. Most subjects of astronomers' cameras are so vast that changes play out over vastly more than human lifetimes, so it is a treat to see things happening in the sky. These wisps varied and even moved over times of a few years. But seeing how they fit together had to await the sharper images of *Hubble*. A 1995 image series obtained for a project by Jeff Hester and Paul Scowen (of Arizona State University) revealed that these wisps were the brightest parts of coherent patterns, with large pieces of tilted rings traced around the pulsar. Furthermore, the changes were associated with these rings and additional narrow jets of matter, streaming directly away from the pulsar. To clarify the nature of these features, and their connection to the pulsar and to the rest of the nebula, Hester orchestrated an unprecedented dual campaign of coordinated observations involving *Hubble* and the *Chandra* X-ray observatory. This campaign is likely to remain unparalleled, since changes in the telescopes' scheduling restrictions no longer allow such extensive coordinated observations.[20] Viewed as a time-lapse movie, the data provide an extraordinary glimpse of processes that are common in the Universe, but usually seen only on larger and longer scales. A single bright ring around the pulsar stands still but flickers, perhaps indicating a standing shock wave where the hot wind flowing from the central pulsar encounters the surrounding nebula. Just outward from it, repeated rings of brighter emission separate and rush outward at 40% of the speed of light. These are plasma waves, in which the energy is carried by interactions between charged subatomic particles which are separated in the energetic maelstrom of the supernova remnant. In addition, the pulsar produces jets of particles which we can see rushing outward and then decelerating in the nebula, just as the enormous jets in radio galaxies are thought to do – if we could watch one over a million years.

The rings marking the plasma waves' passage appear slightly off-center from the pulsar. This is another light-travel illusion. The near side has less distance to travel for us to see it, so we see it in a later point of its expansion than the more distant far side (Fig. 2.26). This makes the ring appear off-centered, by an amount that gives the same $0.4c$ expansion velocity as we measure by looking at the "sides" of the ring pattern.

Over and over, we have encountered astronomical situations in which the time and space transformation of relativity make a difference - sometimes a vast difference – in what we see. As we routinely measure positions and times with greater and greater precision, we need to be correspondingly more careful about our reference frames. As the International Astronomical Union has recognized with successive resolutions defining specific frames of reference we can no longer deal with how we see celestial objects that are "actually" at a certain distance and direction; we must transform relativistically between a target object's natural reference frame and our own.

---

[20] Some of the astronomers involved suspect that this came about precisely because of the trouble caused by scheduling the Crab observations. Regardless, it did make for a unique and revealing data set.

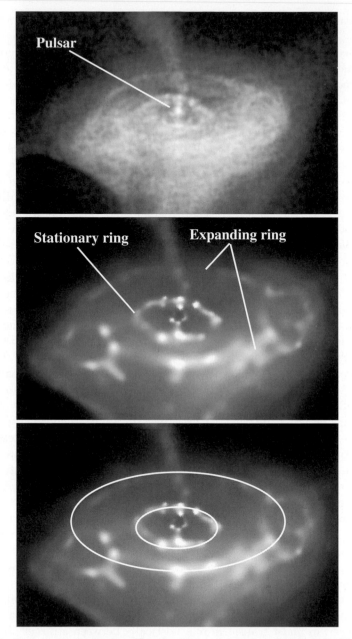

Fig. 2.26. One frame from the *Chandra* X-ray sequence of the Crab Nebula. The top frame shows the X-ray appearance of the nebula, while the middle frame has been processed to make the ringlike structures more visible. The bottom panel superimposes ellipses – one to show that the stationary inner ring appears centered on the pulsar because it is not expanding, and the other to indicate how the outer ring, a wave motion expanding at $0.4c$, appears off-center because of the time delay we see between its near and far sides. The center of the ring appears displaced toward the near side. (Image credit: NASA / *Chandra* X-ray Center / Smithsonian Astrophysical Observatory.)

2.11 Does the Universe lie at the starbow's end?

As great a technological leap as it would require to actually see it, we cannot help wondering what the Universe would look like if we could view it at relativistic speeds (relative, of course, to typical stars nearby). Science-fiction writers have imagined this in various ways, some inevitably hampered by the limited scientific knowledge of their times. Poul Anderson's "Tau Zero" dealt with the time dilation between its passage on board such a vessel and the outside Universe, but it was Frederik Pohl's "Gold at Starbow's End" that put an evocative name to the distortions of starlight we would see. As we moved faster and faster with respect to the stars, multiple effects of relativity would distort their appearance. First, their locations would be shifted forward, by aberration. The geometry of this effect is the same as for tilting an umbrella forward when running through rain, but it takes special relativity to get the amount right for very high speeds. It can be measured telescopically for all stars as seen from the ground, due to the Earth's changing orbital direction (and has been for a long time, having been originally discovered by James Bradley in 1726). The changing compression of the field of view as the orbital direction of the Hubble Space Telescope changes, due to different amounts of aberration at different angles to its motion, sets a limit to how precisely multiple instruments on board can image the sky at the same time. For our imaginary starship, at higher speeds we would see what had originally been more and more of the sky bunched into a small area ahead of us, as our own motion made light arrive from ahead unless it was almost exactly catching up with us.

Further placing most of the action of the bow, stars ahead of us would appear brighter. For approaching objects, the familiar Doppler shift would make all their radiation appear bluer. Relativity shows us that the time transformation between their reference frame and ours makes their clocks run fast compared to ours, so that we might see in a second the light that we would ordinarily see in a minute (just as in the jets of quasars being brightened when they happen to be pointed and moving toward us). The amount of this amplification, and the degree of blueshift, depend on the relative velocity, where the direction as well as speed matters. This realization gave rise to the original description of a starbow – a circular rainbow ahead of the ship, with all the starlight shifted to the blue on the inside and redder on the outside. But, since starlight does not stop at the boundaries of visible light, a starbow would be richer than this. Even when moving at highly relativistic speeds, at each angle ahead of and behind the vessel there would be some kind of celestial object to be seen, its radiation shifted into the visible band. Dead ahead, we might see the cosmic background radiation, pulled up from its microwave murmur to recall the vanished aftermath of creation. And behind us, the flickering X-rays and gamma rays of doomed material near black holes would have lost all but a thousandth of their energy, and remain visible to our eyes as a handful of dim specks. In between, we would see the whole cosmic bestiary sorted as if by an immense prism – the hottest stars on the outside of a broader and more colorful starbow, the coolest red dwarf stars and brown dwarfs inside (Fig. 2.27). The only real gap would be

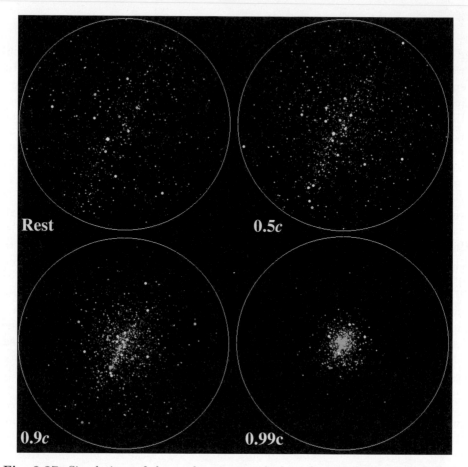

Fig. 2.27. Simulations of the starbow as seen looking forward from the bridge of a starship in the vicinity of Earth, heading toward Orion. Each shows a full hemisphere's view. One image is at rest (that is, our actual view), while the others represent forward velocities of 0.5, 0.9, and 0.99c. As the velocity toward Orion increases, aberration bunches the stars forward at high velocities, and Doppler shifting changes their colors and relative brightness, depending on the broad shape of each star's spectrum. At yet higher speeds, infrared emission from interstellar dust shifts into the visible band straight ahead. (Images courtesy of Jun Fukue of Osaka Kyoiku University, originally produced for the 1991 Japanese-language book *Visual Relativity 2*.) This figure appears as color plate 3.

a ring where we see extreme-ultraviolet light shifted into the visible – a band where interstellar hydrogen absorbs light from all but the nearest stars. Inside the zones of shifted starlight, we might see glowing clouds of interstellar dust at a few tens of degrees above absolute zero, absorbing the feeble starlight of interstellar space.

When we are able to dispatch emissaries at such speeds, the strangeness of the voyage will not have to await landfall on new worlds. It will set in as soon as the vessels embark across the interstellar vastness.

Further reading:

The Web site http://www.astronomy.ohio-state.edu/ãgnwatch/history.html spells out the history of the International AGN Watch.

Much of the history of the discovery of SS 433 is taken from *The Quest for SS 433*, by David H. Clark, Viking Press, New York, 1985.

Felix Mirabel and Luis Rodriguez describe the discovery of microquasars in "Microquasars in the Milky Way", *Sky and Telescope*, May 2002, p.32.

End in Fire: The Supernova in the Large Magellanic Cloud, by Paul Murdin (Cambridge, 1990) describes the discovery and understanding of this uniquely informative event.

A highly personal account of the story of gamma-ray bursts and their association with the collapse of massive stars is given by Jonathan Katz in *The Biggest Bangs: The Mystery of Gamma-Ray Bursts, the Most Violent Explosions in the Universe* (Oxford, 2002).

3 Relativistic Matter

In extreme environments, the effects of relativity change the way matter behaves. Perhaps the least bizarre of these is in white dwarf stars, where general relativity's prediction of the gravitational redshift was borne out (and originally misinterpreted). Neutron stars exhibit much more striking effects from relativity, as light near their surfaces wraps outward on curved paths and particles caught in their magnetic fields are accelerated near the speed of light. Individual subatomic particles at such energies are responsible for most of what radio astronomers see, from cosmic rays and pulsars in our own galaxy to quasars seen in the depths of space and time.

However, relativity does not apply only to such exotic environments. It applies to ordinary matter under ordinary circumstances. The editors of *Time* were being literal as much as poetic when they captioned a cover portrait of Einstein with "all matter is speed and flame". The structure of some kinds of quite ordinary atoms is dictated by relativity, when their internal forces are so strong that the electrons' energies make them relativistic. In the cartoon Bohr atom picture, the electrons are moving close to the speed of light, so that Newton's rules are no longer in control. The most visible example of this is gold. [1] Certain chemical elements have similar properties because they share the same number of electrons in the outermost orbits – one of the main features which led Mendeleev to construct the periodic table of elements familiar in every chemistry and physics classroom. Because the chemical properties of elements depend on bonds involving their electrons, the most loosely bound outer electrons matter the most. Thus, elements with complete sets of electrons "on the outside" do not form strong bonds to other atoms, making them inert gases. In contrast, elements which would have complete outer sets (or "shells") with the addition of one more electron have a strong propensity to attract electrons from other kinds of atoms, making them chemically reactive – including such corrosive substances as fluorine and chlorine, and the ever-versatile hydrogen. By this token, silver (atomic number 47) and gold (79) should share many properties, which they do. Both are fairly rare metals which exist in pure atomic form, and have chemical

---

[1] I first learned this from a character in a science-fiction story by Stephen Baxter, a writer who does his homework in great depth. Another of Baxter's characters describes work done by a radio-astronomer colleague of mine dealing with the arrow of time and fundamental physics, leading that colleague to conclude that more science-fiction authors than astronomers read *Nature*.

bonds weak enough to make the native forms easy to work. But they look different, to the joy of jewellers and customers everywhere. Why the difference? By the time we look at gold, the electrical attraction that its 79 protons exercise on the lowest-energy electrons gives them energies (crudely put, orbital velocities) that are relativistic, shifting their energy levels from simple correspondence with silver. The same patterns that give the broad visible-light reflectivity in silver are shifted to lower energies (longer wavelengths) in gold (which is one of several reasons that infrared-optimized telescopes benefit from having mirrors with a gold, rather than silver, coating). The same thing, of course, happens in all the heavier elements, but the gold/silver pairing is unique in comparing elements which are familiar in pure forms, and have reflective spectral bands sitting conveniently in visible light. When contemplating the Universe, it is worth remembering that many of us carry a little bit of Einstein's relativistic redshift on our fingers.

3.1 White dwarfs

One of the earliest connections between general relativity and the astrophysics of stars came in observations of white dwarfs, later found to be the fading embers of stars which have completed their lives as sources of fusion energy. One white dwarf – the companion of Sirius, or Sirius B – had been known for decades, and in fact had been discovered two more decades before being actually seen. Sirius B was perhaps the first example of "dark matter" in astronomy, having been revealed by its periodic perturbations of the otherwise linear motion of Sirius A early in the nineteenth century. Friedrich Bessel, in Germany, had noticed this change in the motion of Sirius itself in 1840. The dim companion star was actually seen only in 1862, as part of a train of circumstances involving the U.S. Civil War and an administrator who began his career at my own institution. Frederick Augustus Porter Barnard ("Fap" Barnard to his less decorous students[2]) taught physics, astronomy, and mathematics at the University of Alabama from 1838 to 1854, also finding time to make contributions to the photographic process and resolve surveying disputes about state boundaries. Barnard (Fig. 3.1) took the new position of Chancellor at the University of Mississippi ("Ole Miss") in 1856. He soon lobbied the state legislature to appropriate funds for what would be the world's largest telescope, to be located on the campus in the town of Oxford. The observatory structure was in fact built, and today houses the university's Center for the Study of Southern Culture. But war and its aftermath diverted the telescope to a more northerly site. [3]

---

[2] It is to some of these same students, through scribbled rhymes, that we owe evidence that his last name was pronounced with the stress on its final syllable, unlike the more famous astronomer E.E. Barnard.

[3] Barnard had built an observatory at the University of Alabama, housing an 8-inch refractor. It fared somewhat better; the observatory was merely ransacked when the Union Army burned the campus in April 1865, while the objective lenses were safely hidden under the floor at a nearby asylum.

Alvan Clark and Sons, in Massachusetts, had become the premier craftsmen of large telescopes (which at that time meant refracting telescopes, mirror technology not having made the leaps that made reflectors dominant in the following century). Their products became increasing large, precise, and in demand, up through provision of the lenses for the two largest refracting telescopes ever to enter operation – the 36-inch (0.9-meter) instrument installed at Lick Observatory in 1887, and the 40-inch (1.0-meter) telescope at Yerkes Observatory ten years later. When Barnard had succeeded in raising the funds necessary to place an order for the large Mississippi telescope, Clark was the obvious vendor. Despite the dramatic change in circumstances, and after the lenses spent some time in the Clark workshop, the telescope was completed for Dearborn Observatory, which later became part of Northwestern University (Fig. 3.2). Its first discovery came during its initial image tests on actual stars, for which the brilliant Sirius was an exacting target. The younger Alvan Clark (Alvan Graham) inspected the image through the $18\frac{1}{2}$-inch objective lens on January 31, 1862, finding a faint speck of light near the glare of the brightest star in the night sky. Further examination made it clear that this was no optical flaw, but the first detection of the hitherto unseen body long known to be locked in mutual orbit with Sirius.

This was clearly a celestial body quite different from its bright cohort. Their mutual orbital motion indicated masses which were comparable, yet Sirius B shines a mere 1/10,000 as bright as Sirius A. A similar companion turned up around Procyon, the brightest star in the Little Dog, in 1896 (having been suspected from perturbations in Procyon's motion as early as 1844). Its orbit is even tighter than that of Sirius B, so the more powerful 36-inch telescope of Lick Observatory and the eyes of John Schaeberle were needed for its discovery. Since nothing useful was known about the actual sizes of these dim companions, early observers entertained the suspicion that they could be as much like planets as stars, reflecting the light of the primary stars.

A 1915 observation of the spectrum of Sirius B, obtained by Walter S. Adams using the 60-inch (1.5-meter) reflector on Mt. Wilson, showed that its visible spectrum was not dramatically different from Sirius itself, and that it belonged to the same class as a few other faint blue stars such as o^2 Eridani B – perhaps the easiest white dwarf to show visitors telescopically. The scattered light from the bright star has always made precise measurements of the white dwarf difficult, especially in the nights before electronic detectors could allow numerical subtraction of its effects. These were the sorts of measurements that brought out the virtuoso in some observers. Robert Hardie, for example, used the 61-cm telescope of Vanderbilt University for a set of brightness and color measures that were superceded only when telescopes were erected at sites in the Chilean Andes, where Sirius passes almost overhead. His observations required not only excellent atmospheric conditions, but physical modification of the support structure of the telescope's secondary mirror to reduce the effect of the cross-shaped pattern they produced by diffraction of the incoming light. These modifications involved cuts and bends to the edges of the support struts guided by a random-number table, and were successful enough that Sirius B was put on the viewing list for public programs at the university's Dyer Observatory for several years.

Fig. 3.1. Frederick Augustus Porter Barnard (1809-1889), an early advocate of astronomy in American universities. His success in fundraising for a large telescope in Mississippi eventually led to the first sighting of Sirius B, although the telescope was never to leave the North. (University of Alabama photograph.)

Fig. 3.2. The planned and final homes of the Clark refractor which was used to discover Sirius B. Left, the Barnard Observatory in Oxford, Mississippi (courtesy University of Mississippi). Right, Dearborn Observatory of Northwestern University in Chicago, where the telescope has been in operation for well over a century (courtesy Northwestern University).

The faintness of these objects, and the masses implied by their perturbations of Sirius and Procyon, meant that we were dealing with some very different kind of star than astronomers were familiar with. And once their sizes could be estimated from colors (hence surface temperatures) and brightness, the main way they were different had to be enormous density – the Sun's mass packed into a ball the size of the Earth. As has happened so often in our understanding of cosmic phenomena, our understanding of white dwarfs had to await a new understanding in fundamental physics. In this case, it was the Pauli exclusion principle of quantum mechanics, as shown by a young Indian student contemplating the problem en route to graduate study in England.

The brilliant young Indian theoretician Subrahmanyan Chandrasekhar (1910-1995), born in Lahore, went to England to pursue advanced training. Beginning even before his arrival in Cambridge in 1930, he considered the possible role of degeneracy in stellar interiors. Rather than a moral judgment, by *degeneracy* physicists mean a state of matter in which the pressure is determined by quantum-mechanical effects, rather than by the motion of atoms or molecules ordinarily associated with the material's temperature. This arises because of the exclusion principle discovered by Wolfgang Pauli. Within a single quantum system (normally an individual atom), no two particles of certain kinds (including electrons, protons, and neutrons) ever share the same energy state. This principle gives us the chemistry we know, by determining what the electrons' energy properties are and how many will be loosely enough bound to the atom to be free to partake in chemical bonding. If we consider a situation of such density that the "single quantum system" spans many atoms–worth of matter, perhaps an entire stellar interior, all the possible states of the electrons will be filled up to very high energies. Adding one more electron, it would have to have yet higher energy because all the lower levels are occupied. Thus these electrons would always have so much energy that they would yield a pressure in this substance which would be independent of temperature (that is, of the ordinary motion of particles). This so-called degeneracy pressure makes white dwarfs exist; and in a different guise, involving atomic nuclei, neutron stars as well.

Chandrasekhar's supervisor at Cambridge, R.H. Fowler, had published a discussion of the role of the exclusion principle on the density of stellar material, in 1926, with much of the mathematical apparatus worked out. but without making the connection to white dwarfs as visible examples of degenerate matter. Chandrasekhar developed this application, showing that white dwarfs matched the properties of stars in which the electrons exerted degeneracy pressure, and that the energies of some of these electrons were so high that relativistic effects become dominant. This in turn meant that there was a maximum mass that could support itself by degeneracy pressure before gravity would force it to collapse – the *Chandrasekhar limit*, which is about 1.4 times the Sun's mass (depending somewhat on the atomic composition of the star). We will see this mass limit again, since it sets the boundary for stellar remnants between white dwarfs and neutron stars. Chandrasekhar's early work on degenerate matter in stars was a major part of the research cited in his 1983 Nobel Prize.

Not everyone at Cambridge was impressed with this piece of Chandrasekhar's work. Arthur Stanley Eddington – perhaps the most prominent theorist of the time – stated that the idea was so outrageous that "there should be a law of Nature to prevent a star from behaving in this absurd way." Chandrasekhar, barely twenty years of age, received fortunate advice from other senior physicists encouraging him to stand his ground. Despite this, Chandrasekhar's respect for Eddington's intellectual achievements lingered; in 1983, he published a biography entitled *Eddington, the Most Distinguished Astrophysicist of his Time*. As several decades of calculations on the life cycles of stars, confronted with increasing precise data, were to show, stars do behave in this way. A white dwarf is the endpoint of a star whose mass ends up below the Chandrasekhar limit. Once any mass loss from winds and the ejection of a planetary nebula has taken place, the star's starting mass may have been much greater, and certainly all stars below 1.4 solar masses will end up this way.

Eddington had been a champion of the framework of general relativity since his role in the eclipse expeditions of 1919 (Chapter 4). By this time, he regarded relativity's predictions not as ways to test the theory itself, but tools to use for astronomical measurements in testing other, more speculative, theories. This particularly included his own notions about the history of stars and the relations among the various kinds of main-sequence, giant, and white-dwarf stars that present themselves to us. He proposed that the hot interiors of white dwarfs would have stripped atoms of their outer electrons, making the atoms smaller and allowing a denser packing than for neutral gas. He calculated that such an object with the mass of Sirius B would be about the size of Neptune (that is, four times the radius of the Earth), and predicted the relativistic gravitational redshift to be expected on this basis. General relativity tells us that observed time transforms as we look toward or away from matter. Not only do moving clocks run slow, but clocks deeper in a gravitational well likewise run slow. This has by now been seen not only using white dwarfs, and perhaps X-ray spectra of material close to black holes, but on and around the Earth. Gamma-ray spectroscopy using resonances among atoms in crystals (the Pound-Rebka experiment, first carried out at Harvard in 1959) is so sensitive that it can detect the tiny redshift (two parts per quadrillion, 2×10^{-15}) as gamma rays from atomic nuclei are directed downward from the top of a 22-meter tower to detectors at the bottom. This phenomenon was also measured using frequency standards in space, and enters into the calculations for precision navigation using GPS systems.

In this case, the weight of scholarly authority impeded the progress of science. Using the 100-inch (2.5-meter) Mt. Wilson telescope, W.S. Adams finally obtained a spectrogram free enough of compromising light from Sirius itself to attempt a measurement of the white dwarf's gravitational redshift. Waiting for the calmest atmospheric conditions – which on Mt. Wilson are very calm indeed – and using a metal mask to keep as much light from Sirius out of the spectrograph as possible, by 1925 Adams obtained a spectrogram which showed some of the companion's spectrum apparently free of interference by Sirius itself. His derived values for the gravitational redshift[4] was 19 km/s. This value fit well with Eddington's notion – the gravitational redshift expected from something with about the Sun's mass and Neptune's radius.

This value is also wrong. Later measurements with more complete control over error sources – most recently 2004 *Hubble* data analyzed by Martin Barstow and colleagues – gives the gravitational redshift as 80 ± 4 km/s. Fig. 3.3 illustrates how well *Hubble* can separate the two stars and part of the spectrum they obtained is shown in Fig. 3.4. White-dwarf aficionado Jay Holberg has examined this issue closely, and suggests that Eddington's influence was so powerful that Adams decided that the results made enough sense that he did not have to carry out all possible checks for problems. In particular, the spectrum of Sirius A includes many spectral lines not present in the pure hydrogen atmosphere of the white dwarf, some of which could blend with the hydrogen lines and subtly shift their wavelengths without this being visually obvious on the photographic plate. Making matters worse, atmospheric scattering of light increases for bluer light, so Adams was restricted to using only two strong absorption lines in the red and blue-green regions of the spectrum; the hydrogen atmosphere of Sirius B produces no other features to measure, although Adams could not have known this. The lesson here is that observational results which make perfect sense are no less in need of cross-checks and scrutiny than are those results that do not fit with anything else we know. Adams may have felt no need to investigate problems of scattered starlight any further, since his value was in accord with the best (or, at any rate, most widely trumpeted) theoretical notions of the time. This mistake could only be uncovered forty years later, when Sirius B's orbit brought it once again farther from the glare of Sirius A and allowed relatively uncontaminated observation with Earthbound instruments.

Only for a few white dwarfs has this gravitational redshift been measured, because it is the special circumstance of lying in a well-resolved binary system that lets us separate this modest shift from the Doppler shift due to the radial velocity between observer and star. Following the white dwarf and its companion star through a complete orbit (or enough of it for an accurate orbital determination, at least) will tell us the mean velocity, which must be the same for both members of a binary pair. It is the excess redshift for the white dwarf that we can attribute to the transformation between its surface reference frame, deep in its gravitational distortion of spacetime, and our frame, which by comparison is practically in empty flat space. When we can separate the Doppler and gravitational shifts, it furnishes a powerful way to "close the loop" on our understanding of these objects. We have the star's mass from the relative orbits of white dwarf and "normal" companion, and an estimate of its radius from models of the stars' atmosphere fitted to its spectrum (more or less following

---

[4] Redshifts are often quoted for convenience in terms of the velocity that gives the same shift from the Doppler mechanism, although it is evident in this case that such a velocity is not involved. It is easier to remember few-digit numbers than the z value, which is $z = 0.000267$ for this white dwarf from modern data. However, the velocity units should not be taken to imply an actual surface motion of the white dwarf, much less a difference in the speed of light received from it.

Fig. 3.3. The crisp images from *Hubble* clearly distinguish the faint white dwarf Sirius B from its brilliant neighbor. This image, taken at the near-infrared wavelength of 1 μm by the telescope's Wide Field Planetary Camera 2, shows the white dwarf to lower left. The observation was deliberately designed with a mode giving the bright multiple diffraction spikes from Sirius A so as to get the most accurate positional measurements relative to the bright and saturated core of its image. This image, from October 2001, was kindly provided by Howard Bond, Space Telescope Science Institute (NASA).

the blackbody radiation law, which relates radiated energy to temperature and surface area). Given these values, general relativity predicts the gravitational redshift, so to the extent all these values match, we might consider our understanding of white dwarfs (and relativity) to be on the right track. Increasing the number of binaries with such data has led to repeated *Hubble* images of some of the brightest stars, to track their dim white-dwarf companions free of the complicating effects of atmospheric blurring. *Hubble* is also a powerful platform for this because the images can be taken well into the ultraviolet, where the white dwarfs are brighter, and the "normal" stars are generally fainter, than in visible light. At the extreme, images of Sirius in the soft X-ray band are thoroughly dominated by the surface emission of the white dwarf at a temperature of about 25,000 K, reversing the visual appearance in which the main-sequence star Sirius A outshines its companion by 10,000 times.

3.2 Neutron stars

A species more outrageous than white dwarfs swam into our ken first as a purely theoretical concept. Only a few years after the neutron was discovered in the

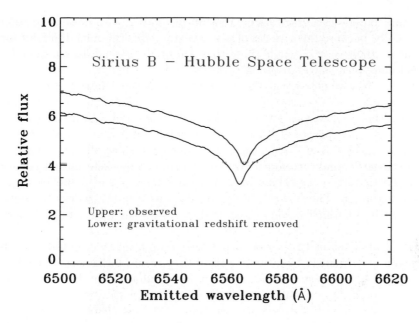

Fig. 3.4. Part of the spectrum of Sirius B around the deep-red Hα absorption line of hydrogen, illustrating the gravitational redshift. The upper curve plots the observed spectrum, while the lower one shows the spectrum that would be seen without the gravitational redshift (but including the orbital motion of Sirius B in its binary system and the binary's motion relative to the Sun). The shift beween the two shows the redshift of light emerging through the curved spacetime around the white dwarf. This observation did not tax the light-gathering power of *Hubble*, requiring only 10.5 seconds' worth of exposure, but did rely on its exquisitely sharp images to keep the light of Sirius A from contaminating these results. Data from the NASA/ESA Hubble Space Telescope archive, original principal investigator Martin Barstow.

laboratory, Walter Baade and Fritz Zwicky considered the properties of a star in which not merely the electrons, but also the neutrons normally confined to atomic nuclei, were so tightly packed as to become degenerate.[5] Such an object would be tiny by astronomical standards, with a radius of perhaps 15 km containing 2–3 times the Sun's mass. These configurations could be understood only by using quantum mechanics modified for the energy behavior specified by special relativity. Not only would the energies of neutrons be relativistic, but the depth of its gravitational well induces decidedly non-Euclidean behavior to light emerging from or passing close to its surface. Compared to a gravitational redshift of $z = 0.00026$ for the white dwarf Sirius B, light from the surface of a

---

[5] And in producing such an object, the pressure would become high enough to force electrons into the nuclei, where they can combine with protons to yield more neutrons. This process also releases energy carried by neutrinos, one of the processes that drives supernova explosions.

neutron stars undergoes a gravitational redshift of $z \approx 0.2$. This does us no good observationally, because the surfaces of neutron stars are still much too faint for us to measure their spectra, while those that are bright enough to see shine by processes taking place far above their surfaces.

Baade and Zwicky noted that the gravitational energy released, if one considered collapsing a massive star into a neutron star, would suffice to power the brilliant supernova explosions which Zwicky had recognized as a distinct class, and suggested that supernovae did in fact accompany this final transition of a star. Energy can be released when an object contracts, since all parts of it become closer together and therefore more tightly confined by their mutual gravity (or in relativistic terms, the curvature of spacetime increases at points closer to the generating mass). The energy which can be released in this way is the same as the energy input which would be required to return the object to its former size.

Neutron stars remained a theoretical curiosity for almost three decades. Even if they began at extremely high temperature, they would be almost impossible to detect because they have so little surface area to radiate from. Even a nearby neutron star would shine at something like 27th magnitude, far beyond the reach of photographs even with the Palomar telescope. The best chance of eventually seeing one was thought to be by looking for the X-rays given off by an intensely hot surface, although this, too, appeared to be a task for future technological generations.

Some neutron stars did not have to await a further generation of satellite X-ray telescopes to reveal themselves. The story of the discovery of pulsars, late in 1967, has been told often, not least in introductory astronomy classes to demonstrate the human issues which can be involved in science. It begins with radio astronomers' persistent dissatisfaction with the level of detail their telescopes show them. Even as interferometry between separate antennas was being developed, astronomers had learned to use interplanetary weather to estimate the sizes of radio sources, in the same vein as the general rule that we see stars twinkle, but not the much larger images of planets. In both cases, this works because we see distant objects through a turbulent and irregular foreground medium, which will refract the incoming radiation in changing ways. A very small object (loosely, one appearing smaller than the typical size of changes in the refracting medium) will brighten and dim erratically as some of its radiation is refracted into or away from our line of sight. For larger objects, the effect is muted because the effects on different parts of it will average out. In radio astronomy, this is known as interplanetary scintillation, and happens as the ionized solar wind blows outward past the Earth. This effect is most pronounced for the longest-wavelength radio waves, which are in just the ones for which our telescopic vision is blurriest. The payoff for being able to estimate even average sizes for radio sources could be immense, since we appreciated very quickly that radio surveys were penetrating space to such great distances that we could expect to see radio sources evolve with cosmic time.

This was the scientific background that brought Cambridge graduate student Jocelyn Bell to wire up 4.5 acres of English countryside with radio antennas.[6] Working at a wavelength near 4 meters, these antennas were designed to scan the piece of sky passing the meridian, thus avoiding the enormous expense of building a similarly large piece of moveable equipment.[7] Besides, in studying the deep Universe, one piece is as valuable as any other as long as nothing is blocking your view (a philosophy both guiding and supported by the optical and X-ray "deep field" observations which have proven valuable in the past decade). Once working, the telescope recorded the radio signals passing through its beam, generating paper for analysis at the rate of 29 meters per day.

By late 1967, Bell had noticed the sort of signature in her data that gives astronomers the odd feeling that either a major discovery is at hand, or (more often) something quite unexpected is compromising the whole experiment. At various times, the chart recordings would show "scruff" in one of the telescope fields – sharp, repeating flashes of radio emission, recurring for several minutes with a period of 1.3 seconds. This looked, at first, like the radio interference that drives radio telescopes to remote locations and radio astronomers to ever more sophisticated techniques of analysis. Still, none of the obvious kinds of interference would match this behavior – especially its recurrence when the same piece of the sky rotated into view each day, rather than at the same clock time. In late October 1967, Burnell went to the telescope when the mystery object should be crossing the field of view and switched the chart recorder to a higher speed, to show the flashes in more detail. Her thesis advisor, Anthony Hewish, used these data to establish that the flashes recurred from the same region of sky, thus eliminating most possible sources of interference, and involved more colleagues in observations to establish the distant nature of the source. Controversy has continued to cloud the recognition for the discovery of pulsars. Hewish received a share of the 1974 Nobel Prize in Physics, with a citation featuring his role in this discovery. Many astronomers thought this selection unfairly slighted Bell (now Bell Burnell, for those seeking more information), although Hewish's Nobel lecture made no secret of the importance of Bell's persistence and dedication. As she has written, "demarcation disputes between supervisor and student are always difficult, probably impossible to resolve."

Their data quickly brought to light several additional pulsars. These showed that such pulsing sources were common enought to most probably *not* be the initially plausible "Little Green Men", and provided further data to narrow the range of possibilities. The data were sparse indeed. We were dealing with some kind of astrophysical object, capable of doing something or other (rotating, pulsating, nonradially oscillating) on a very regular schedule in a second or two, and giving off lots of energy in the form of radio waves every time it did so.

---

[6] Since appropriate credit matters in this story, she has written that she was one of perhaps five students involved in the construction, plus several cheerfully hammer-wielding summer students, although Bell was solely responsible for the operation of the equipment.

[7] Careful use of delays in the receiving electronics let them actually scan four such strips of the sky, separated north–south.

There were calculations of how fast white dwarfs could spin (not fast enough) or pulsate (just fast enough), even neglecting the question of what would make them radio sources. And the notion of neutron stars recurred – in some cases early enough for prestigious journals to reject this idea for publication on the grounds that it was complete speculation. [8] At least the flashing behavior gave us a name to help organize our ignorance – by analogy with stars and quasars, these objects were called pulsars.

Other observatories quickly became involved in pulsar studies, especially surveys to determine how many of these were within our range and where they were located. The 300-foot (91-meter) radio telescope of the National Radio Astronomy Observatory in West Virginia turned up a pulsar coincident with the Crab Nebula, whose period was refined by a team using the giant 305-meter Arecibo telescope. The Crab pulsar's period was only 33 milliseconds, flashing 30 times a second – about the rate at which a TV set or computer monitor refreshes. At one stroke, this discovery eliminated white dwarfs from consideration, and linked pulsars to supernova remnants (a link strengthened when a pulsar was soon uncovered in the Vela supernova remnant). White dwarfs can neither oscillate nor rotate so fast without breaking up. This gave renewed vigor to the suggestion, championed perhaps most forcefully by Cornell's Thomas Gold, that pulsars are rotating neutron stars whose magnetic fields dictate that radiation emerge in narrow cones, sweeping the sky like lighthouse beams as they spin. Due to its immense density, a neutron star can rotate at thousands of times a second without breaking up. The Crab and Vela pulsars strengthened the case, occurring at the sites of supernova only 1,000–10,000 thousand years ago. Baade and Zwicky had proposed that neutron stars form during some supernova explosions, and these pulsars were the fastest yet found and thus plausibly younger than average. Not only could we actually detect (some) neutron stars, but their radio pulses acted as very effective public-relations services.

The general picture we have of pulsars begins with the end of a massive star's life. As the star fuses ever heavier elements in its core, each successive fusion process yields less energy, until the core is dominated by iron nuclei. Iron occupies a special place among the elements, as no nuclear process involving iron produces more energy than it uses. Neither fusion nor fission of iron nuclei is a net energy source. So, when the star has formed an iron core, there is no longer enough energy being generated to hold the star up against its own gravity, driving the rapid collapse of the star's core under the unimaginable weight of its outer layers. Within milliseconds, the former core of the star is crushed into a neutron-rich ball, emitting vast amounts of energy in the form of neutrinos as protons and electrons are forced together into neutrons. This core is now essentially a neutron star, although a very hot one. Degenerate matter is very

---

[8] This is an instance of a law I have observed in scientific publication: there is no advantage whatever to being more than three years ahead of your time. When the time comes, your paper will be forgotten and someone else will be widely cited. This matters only because citation tends to influence future allocation of research resources.

"stiff", forming a wall that reflects much of the infalling material and driving a shock wave outwards through the star. This material is so dense that it absorbs energy from the neutrinos, helping drive the shock wave outward and blowing the star apart in a supernova explosion. [9] This much of the scheme received dramatic support with the measurement of neutrinos from Supernova 1987A in the Large Magellanic Cloud. After such a cataclysm, we would see the fading glow of the debris, flying away from the former star at thousands of kilometers per second and gradually cooling over the years.

Meanwhile, the neutron star, the crushed former core of the star, has maintained part of the star's magnetic field within it during the collapse. Confined with a volume perhaps millions of times smaller than it started with, the magnetic field is correspondingly stronger. Any misalignment it had with the star's rotation axis is amplified during this process, so we might see a young neutron star with magnetic poles aligned near its equator and a field a trillion times stronger than the Earth's, in a configuration wobbling around with every rotation of the neutron star. As time goes on, the pulsar slows down, losing energy to its surroundings through the magnetic field, and its pulses also fade. We see only a fraction of active pulsars – those whose beams happen to sweep over us as they spin. And we do not see inactive pulsars at all unless something happens to spin them up and drive their beams once again.

The interaction of pulsars with their surroundings is dictated by relativity. The magnetic field of the neutron star is firmly anchored in its dense neutron-degenerate matter, whipping the entire magnetic pattern around with each rotation. Any charged particles in the vicinity – protons and electrons alike – will be locked to the magnetic field lines and likewise whipped around. But, at the high rotation speeds of pulsars, this will not work for the whole magnetic field. There will come a radius, at the *light cylinder*, where the particles would have to be carried around at the speed of light. As we have seen in Chapter 2, it takes more and more energy to accelerate particles near the speed of light, where they behave as if their mass increases without limit. This acts like a drag on the magnetic field, conveyed back to the spinning neutron star. Particles on lines of the magnetic field well inside the light cylinder can be trapped in configurations that shuffle around without net effect, but the field lines cannnt cross the light cylinder, and they remain "open", flinging particles outward at relativistic speeds. Outside the light cylinder, the magnetic field lines are wrapped into tight spirals. A particle at a fixed point in this region would experience a periodic change in electric and magnetic fields as these lines sweep by once per rotation period, equivalent to electromagnetic radiation of frequency equal to the pulse period. Such extremely-low-frequency (ELF) radiation is absorbed very efficiently in plasma, heating the surroundings and driving the hot winds seen around young, fast pulsars. This heating is where most of a pulsar's ro-

---

[9] In comparison with the cosmologically useful Type Ia supernova explosions discussed in Chapter 8, this scheme applies to Type II supernovae and with modifications to the star's previous history, Types Ib and Ic.

tational energy is sapped; the excellent timekeeping stability of pulsars lets us watch as they slow down from year to year.

While it furnishes us a good description of the pulsar's history, this part of the scheme does not tell exactly where the radio pulses come from. A further part of the standard description of pulsars does that; this radiation is such a small part of the overall energy budget that its details remain uncertain. As pulsar observer Don Backer has written: "While this standard model for radio emission is incomplete, and may be erroneous in parts, it serves a useful function by describing the wealth of pulsar properties." Near the magnetic poles, which may be aligned almost anywhere with respect to the rotation poles, charged particles will be accelerated almost straight outwards, to such high velocities that relativistic beaming is important. Particularly when these field lines cross the light cylinder, they will be accelerated to very high velocities, producing very narrow beams. Some pulsars show very detailed structure in these beams as they wash past us, suggesting that they begin in regions only a few hundred meters across near the pulsar poles. Such deductions represent truly remarkable discrimination of detail across interstellar distances.

A handful of young pulsars have particle beams so energetic that we see the flashes not only in the radio, but in the optical and right across the electromagnetic spectrum. The Crab pulsar was conclusively identified by its pulsed optical emission (Fig. 3.5), a feature which continues into the ultraviolet, X-ray, and even gamma-ray regimes. Astronomers at several observatories were searching for such a signal, using high-speed electronics to capture the output of a photoelectric tube. Here, the race went not to the first to observe, but to those whose technique gave an answer in real time. John Cocke and Mike Disney, using a 0.9-meter telescope on Kitt Peak, saw pulses emerge on an oscilloscope synchronized to the pulsar period and knew they had a detection, while R.V. Willstrop in the UK had taken data months earlier but was still engaged in a tedious and time-consuming reduction process. Soon thereafter, observations with a photomultiplier on the Kitt Peak 2.1-meter telescope had narrowed the position to a single "star" in the center of the nebula, dramatically confirmed by TV images using a strobe disk at Lick Observatory, by two guys named Joe.[10] The moment of discovery by Cocke and Disney happened to be recorded on audio tape, an unusual gift to posterity.

The rotation periods of pulsars are marvellously stable, making them the best clocks in astronomy. This stability has allowed surprising discoveries. The first planets found outside our Solar System orbit, of all things, a pulsar, presumably the remnant of a supernova from whose debris a born-again planetary system emerged, well off toward the Galactic Center. Binary-encoded pulsar periods were used as calendrical markers for any alien finders of the *Pioneer* and *Voyager* probes, with their plaques marking their time and place of origin. And pulsars in binary systems provide the only specific evidence to date for the gravitational radiation predicted by general relativity (Chapter 9).

---

[10] Wampler and Miller, respectively.

Fig. 3.5. The optical flashes of the Crab Pulsar, captured in a series of time-sliced images strobed to its 0.03-second pulse period. In successive frames, spaced one millisecond apart and coadded across two hours – worth of observation, both the bright main pulse and a fainter secondary pulse can be seen. A random field star to the pulsar's upper left provides a reference point for each subimage. These images were produced using the 4-meter Mayall telescope of Kitt Peak National Observatory and a time-tagged imaging detector. (Courtesy of Nigel Sharp.)

Most neutron stars are not pulsars – certainly not active and observable ones. Turning back to the first predictions, only recently have we been able to see an isolated neutron star which does not produce radio pulses, using first X-ray sources which did not have a brighter optical counterpart, followed by the high resolution and ultraviolet sensitivity of *Hubble's* cameras (Fig. 3.6), finding what may finally be the bare, hot, and bizarre surface of a neutron star. The combined X-ray and ultraviolet measurements indicate a surface temperature of 1.2 million K.

Neutron stars are the likely remnants of most supernova explosions. As for white dwarfs, neutron stars have an upper mass limit. This one, though, is less well defined – not only because of uncertainties in our understanding of matter under these conditions, but because their magnetic fields can be so strong and their rotations so rapid as to provide significant support against gravity. The nominal upper mass limit is around 2 solar masses, but with the possible effects of rotation and magnetism, astronomers have been more likely to adopt a value of 3 solar masses to be sure of distinguishing possible black holes from neutron stars.

3.3 Atom smashers and radio telescopes

The sky revealed by radio telescopes is a very different one from the familiar starry sky of optical astronomy. The visible Universe is dominated by stars – which is no surprise, since our eyes need to make best use of the light from the nearest one. Very few stars are radio sources. Instead, the view features diffuse material in distant galaxies and between the stars of our own Galaxy (Fig. 3.7). We see the expanding bubbles of exploded supernovae, and the similarly expanding clouds of matter whch betray the action of enormous black holes in other galaxies. Most bright radio sources lie beyond our galaxy, and many have a characteristic double pattern on either side of the visible-light galaxy (Fig. 3.8).

The nature of this radio emission became a puzzle almost as soon as Grote Reber brought it to the attention of traditionally-trained astronomers. Experience with stars made them think of blackbody radiation, matched to an object's temperature. But the behavior of the radio emission was very different from a

Fig. 3.6. Hubble Space Telescope image showing the dim glow of an isolated neutron star (as the faint dot at center). Comparison with X-ray data indicates a surface temperature slightly above 1,000,000 K. (F. Walter, SUNY Stony Brook, and NASA.)

blackbody at any temperature, spanning over a decade of frequency with a single, smooth curve gradually rising to longer wavelengths and lower frequencies. This was clearly *nonthermal radiation*, whose nature had to be understood if we were to make any progress in learning about this new window on the Universe.

As astrophysicists eventually learned, most of the sources picked up by radio telescopes are seen because they give off *synchrotron radiation*, otherwise (if not more helpfully) known as magnetobrehmsstrahlung. Electromagnetic radiation is given off whenever particles with an electrical charge are accelerated, either by the changing currents in a radio antenna, interactions with other particles, or by encountering a magnetic field. Particles with the least mass for a given electrical charge are accelerated most strongly and give off radiation most effectively, so we can usually deal only with the electrons in this context (omitting the accompanying protons and heavier atomic nuclei). If we trace the movement of an electron through empty space where a magnetic field exists, it will spiral around the lines of the field. If it starts out moving perpendicular to the field lines, at low velocities, its motion will be a circle. This is analogous to a cyclotron's pumping energy into a beam of particles, with magnetic fields designed

Fig. 3.7. Synchrotron radiation dominates our sky when viewed at radio wavelengths, as seen in this wide-field survey from the Bonn 100-meter and Parkes 64-meter radio dishes. The plane of the Milky Way is prominent across the center, with loops from ancient supernova explosions stretching upward and downward. The nearby radio galaxy Centaurus A, with its twin lobes of radio-emitting plasma, appears above the galactic plane at right. This image, compiled from observations at wavelengths near 70 cm, spans 180° of sky. (Data from the Max-Planck-Institut für Radioastronomie, Bonn.)

to speed them on their way every time they come around in a circle. Once the particles are moving at a fair fraction of the speed of light, the magnetic fields of a cyclotron will become progressively off-kilter from the particles, which do not accelerate as rapidly for a given input, as their effective mass increases. To keep the process going, the magnetic field must increase in synchrony with the increasing energy of the particles, as the relativistic relations between energy and velocity take over – and we have a synchrotron.

In space, an energetic particle may trace a long helical path through regions of magnetic field. It gives off energy continually, due to its acceleration into this

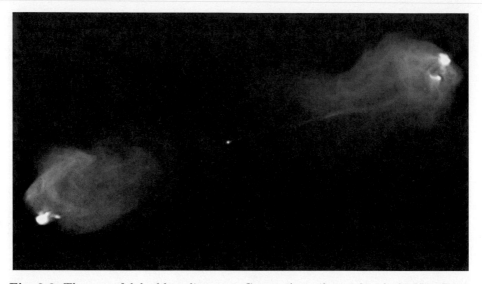

Fig. 3.8. The powerful double radio source Cygnus A, as observed with the Very Large Array at a wavelength of 6 cm. This high-definition image shows not only the twin lobes of emission, but the core associated with the active nucleus of the galaxy itself, and some parts of the jets connecting this core to the lobes about 250,000 light-years from it. These observations show some of the complexity of these lobes, driven by structure in the magnetic field as it controls the synchrotron radiation from relativistic particles. Data from from the National Radio Astronomy Observatory CD-ROM compilation "Images from the Invisible Universe", as published by R. Perley, J. Dreher, and J. Cowan in 1984.

curving trajectory, and it does so – it loses energy and gradually slows down. From the particle's point of view, it gives off the energy in a broad doughnut-shaped pattern perpendicular to its acceleration. But, since the particle may be moving within 0.001% of the speed of light, an outside observer not sharing this motion will see something quite different. From our point of view, the radiation will be beamed (Fig. 3.9) – compressed into a narrow cone ahead of the particle, narrower for higher energies. [11] We therefore see each particle only when it is approaching, and during that piece of each circle of the helix, we see its time compressed by this motion and therefore see its radiation enormously blueshifted and brightened. The effects of beaming and time transformation are convenient to separate for explanation, but are both part of the relativistic transformation from the particles' reference frame to ours.

Similar radiation comes from particles at low velocities, concentrated at the frequency of the orbital motion and its harmonic values. We do see such cyclotron

[11] The angular radius of this beaming cone in radians is roughly the inverse of the Lorentz factor. For example, an object approaching us at $0.9c$ has a Lorentz factor of 2.3, so its radiation appears in our reference frame to be beamed into a conical area ahead of it with a radius of 0.4 radians or $24°$. For an object at $0.99c$, the cone's radius shrinks to $8°$.

radiation from particles near pulsars and other compact objects, as long as the particles are moving at non-relativistic speeds. Adding up the contributions from all the particles at a given energy, synchrotron radiation does not produce a single spectral line or a set of a few, but produces instead an extremely broad smear across orders of magnitude in frequency, which will be further increased in real situations by the different energies that particles will have (to say nothing of the magnetic field changing from place to place).

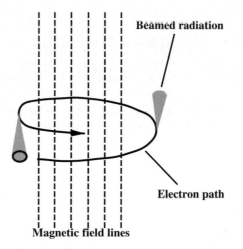

Fig. 3.9. The production of synchrotron radiation. The helical curve marks the path of an electron moving upward through the magnetic field, being continually accelerated at right angles to this drift. At relativistic energies, not only does an electron give off radiation as the field accelerates it, this radiation is tightly beamed forward at each point on its path. From our vantage point, the radiartion we see is dominated by the electrons whose beams point in our direction (as highlighted in black).

After half a century of sporadic mentions in the theoretical literature, synchrotron radiation was first observed in the laboratory in 1947, by a group working at the General Electric development lab at in Schenectady, New York, under the leadership of Herb Pollock. The only synchrotron in operation before GE's was a unit in the United Kingdom, whose designers missed this discovery only by virtue of selecting an opaque rather than transparent vacuum tube at a key location. For particle physicists interested in what can be learned from the results of slamming particles together with ever-higher energies, synchrotron radiation is a nuisance. The power radiated in this form saps the very energy that the physicists have gone to so much trouble to impart to the particles. However, it has seen interesting laboratory use in its own right, so widely that the US Department of Energy funds a National Synchrotron Light Source at Brookhaven National Laboratory, to be a purveyor of intense, well-collimated radiation across a broad span of the high-energy spectrum. The beams extend well into the X-ray regime, giving a selection of wavelengths and wavelength span unachievable by lasers. Detailed analysis of the way the synchrotron beam interacts with matter has been important not only in measuring basic properties of atoms, but in studying the properties of materials on atomic scales (which has become a major industrial driver) and in analysis of biochemically important molecules.

Astrophysicists' understanding of synchrotron radiation – how to identify it, where it comes from, and how it affects the objects they observe – migrated from the rather different community of people who studied cosmic rays, a category of relativistic matter in its own right.

3.4 Cosmic rays – particles with punch

Cosmic rays were identified in the early days of study of radioactivity, when Austrian physicist Victor Hess showed that electrically charged particles encountered his equipment (an electroscope) more frequently as he ascended, passing 5 km altitude during a 1912 balloon flight. Hess reasoned that the source of such particles (at that time not clearly distinguished from genuine electromagnetic radiation) must lie at still greater altitudes, and this discovery was rewarded with a Nobel Prize in 1936 (one of many awarded for work on cosmic rays). By this time it had been found that these "cosmic rays" were affected by the Earth's magnetic field, so they must be particles of matter, and electrically charged particles in particular rather than X- or gamma rays carrying similar energies. This gives cosmic rays of various velocities and masses complex and varied trajectories near the Earth, and caused considerable confusion in the early years because of the very different magnetic latitudes of Austria and Texas; lower-energy cosmic rays do not penetrate near the Earth's surface in the Lone Star State. In fact, a similar role is played by the Sun's magnetic field, whose influence expands and contracts with the cycle of solar activity. At sunspot maximum, we are protected against higher-energy particles than at sunspot minimum. This may be reflected in terrestrial weather patterns; there is some evidence that precipitation can be triggered by cosmic-ray particles, one indirect way in which solar activity may influence our climate.

For several decades, physicists were interested in cosmic rays not only as probes of the Universe, but because they were abundant sources of particles with energies that laboratories could not yet match. During this period, the best way to do certain experiments in high-energy physics was to prepare the target material, take it to a mountaintop or put it on a balloon, and wait for it to be hit by an appropriate particle from space. Indeed, the first detections of the anti-electron – the positron – and of the mu meson happened thanks to cosmic rays. Most of the cosmic-ray particles detected near the Earth's surface are so-called secondaries – particles created from the enormous energies of some of the primary cosmic rays, when they encounter atoms in the upper atmosphere. Some carry enormous amounts of energy, as they move only fractionally slower than light, so that this energy-to-mass transformation can yield showers of vast numbers of particles and antiparticles, some so energetic that they in turn collide with additional atoms to create more particles in a wide-ranging cascade. Perhaps the first record of the primary cosmic rays was obtained by US Army Captains Albert Stevens and Orvil Anderson, who carried photographic emulsions on a world-record balloon ascent in 1935, reaching an altitude of 20 km.

Cosmic rays have been measured with successively more sophisticated equipment – photographic films, cloud chambers in which the passing particles induce condensation trails, Geiger counters, and eventually all-electronic systems which can be carried above the atmosphere to be imbedded in the primary cosmic radiation. Satellites uncovered the best-known effect of cosmic rays, the Van Allen belts of particles trapped in regions surrounding the Earth. The more massive protons penetrate deeper in the Earth's magnetic field and dominate the inner belt, while electrons gyrate around the magnetic-field lines in the outer belt.[12] The Van Allen belts were the first spectacular scientific discovery of the Space Age, recognized by James Van Allen in 1958 based on radiation detectors installed on the US satellite *Explorer 1*. Their properties at higher altitudes were sampled over the following year by *Pioneer 1–4*, in the course of their failures to reach their intended destination of the Moon. Ironically, they could have been detected by instrumentation on board the earlier Sputnik II, and would have been known by a different name, but the Soviet program at the time employed only real-time transmissions with the satellites, and the early Sputniks passed over tracking stations in the the USSR at points in their orbits outside the radiation belts. Since this discovery, James Van Allen has remained a towering figure in space science for four decades, and for much of that time has been a vocal critic of human spaceflight, seeing it as irrelevant for scientific purposes. Thus, as veteran space observer and two-decade Internet legend Henry Spencer points out, "the Van Allen belts – a deadly hazard to manned spaceflight, and a serious nuisance to some classes of unmanned missions – are the perfect namesake."

These zones form a hazard to be reckoned with for human and robotic space explorers alike. All Earth-orbiting piloted missions have stayed below the inner belt (and the moonbound Apollo astronauts flew through them quickly enough to avoid significant health hazards, although they did report seeing flashes of light which were identified as cosmic-ray particles interacting with retical cells), and the operating strategies for low-orbiting satellites must take account of their exact structure. The Earth's magnetic field is not quite like that of a giant bar magnet, having various irregularities and lumps. One such feature brings the inner Van Allen belt unusually low over the South Atlantic, in a structure known as the South Atlantic Anomaly (SAA). Such satellites as *Hubble* and the *Far Ultraviolet Spectroscopic Explorer* (FUSE) are commanded to turn off the most sensitive electronics during orbital passages through the SAA. Otherwise, particle passages would produce spurious electrical voltages, and almost no useful data can be obtained in any case.

The inner radiation belts are also open to human interference, as was demonstrated in the early 1960s when both the US and USSR conducted nuclear tests

---

[12] Physicists sometimes distinguish a weak third van Allen belt inside of these two, consisting of heavier atomic nuclei that reach equilibrium still deeper in the magnetic field

in space.[13] These tests were intended not so much to probe the behavior of the radiation belts as to understand the effects of potential nuclear antimissile weapons. The power of electromagnetic pulses was properly understood once the July 1962 *Starfish Prime* device was exploded 400 km over Johnston Island in the central Pacific. This 1.4-megaton bomb accidentally blacked out power in Honolulu, 1,000 km from its launch site. These tests also temporarily boosted the particle population in low Earth orbit. At least three satellites (of the few yet in orbit) failed as a result, with the major damage being loss of power as the solar cells were degraded by particle impacts over a period of weeks. The enhanced radiation belts were a major concern for spacefaring, since the elevated particle levels could persist for weeks (and at higher altitudes, possibly several years). This came to the fore in July 1962, when the USSR convened a state commission to investigate the safety of cosmonauts Andrian Nikolayev and Pavel Popovich as they prepared for the back-to-back launches of *Vostoks* 3 and 4, a month after the US exploded *Starfish Prime*.

Even as terrestrial particle accelerators could exceed the energies of all but the rarest cosmic rays, understanding these visitors from space remained important. These particles had somehow picked up enormous amounts of energy, flying across interstellar space at speeds only fractionally slower than light itself. They must be telling us about some of the most energetic processes in the cosmos – if only we knew what sort of environments produce them, and how. They must originate from environments in which vast amounts of energy can be imparted to individual particles and atomic nuclei, either through the intensity or sheer scales of the processes. Most analysis has centered on the Fermi process. This was proposed by the Italian–American physicist Enrico Fermi, the "Italian navigator" who first demonstrated a fission chain reaction beneath the Chicago stadium. Fermi is legendary among physicists for his equal prowess in theory and experiment, and played a key role in the Manhattan Project. He noted that charged particles could be reflected back and forth between an interstellar shock wave and the surrounding rarefied material, gaining energy magnetically at each reversal. Eventually, the particle would escape the shock when it acquired too much energy for the magnetic fields to curve its path back into the shock wave. Geophysicists have been afforded a glimpse of this process where shocks form as the solar wind impinges on the Earth's magnetic field, providing most of the particles feeding the Van Allen belts and the displays of aurorae. The Fermi process is appealing, not only because it *must* be operating wherever shocks penetrate the interstellar medium, but because calculations show that it would produce the sort of wide distribution of particle energies that we see. Shock waves arise from the explosive blasts of supernovae, and less violently from encounters between interstellar clouds in slightly different galactic orbits.

It has proven challenging to pin down individual sources of cosmic rays, because their directions of travel corkscrew through the solar wind and terrestrial

---

[13] Two each by the US and USSR during the Cuban Missile Crisis alone, which is at best a testimony to the power of bureaucratic inertia and the failure of memos to reach their intended recipients.

magnetic field before we detect them. Even worse, they must have been deflected many times by interstellar magnetic fields before arriving in the Solar System. A typical energetic cosmic ray swerves though the Milky Way for perhaps 30 million years before escaping the galaxy altogether.

These energetic particles will become less energetic as they encounter interstellar magnetic fields. Each deflection of their travel requires an acceleration, and accelerated charged particles produce electromagnetic radiation. It was in this connection that Vitali Ginzburg and S.I. Syrovatskii worked out the details of synchrotron radiation, first appearing in their book *The Origin of Cosmic Rays*. Their results became known to Western scientists through a pair of review articles translated for publication and still widely cited 40 years later. Through its history of gradual energy loss, a particle emits progressively lower-energy radiation, so that when we look at a whole population of particles, we see radiation spread across an enormous range of frequencies. The nearly scale-free distribution of *particle* energies produced by the Fermi process will give rise to a linked distribution of *radiation* energy.

Many of the "cosmic rays" detected near the Earth's surface are secondary – produced in vast showers when a primary particle collides with a particle in the upper atmosphere. Large numbers of particles and antiparticles can be created in such collisions, forming an air shower similar to that produced by the most energetic gamma rays (section 2.8). These air showers furnish a classic example of the relativistic dictum that "moving clocks run slow". These showers of particles include large numbers of muons – lightweight particles which, left to themselves at rest, decay with a half-life of only two millionths of a second. In Newtonian physics, we should not see these so far from their high-altitude production sites. Even at the speed of light, a muon would travel only about 600 meters before having a 50% chance of decaying. Instead, we detect significant numbers tens of kilometers below their points of origin. This can occur because they are produced with velocities close to c – the transformation from the muons' reference frames in which they do have a half-life of 2 microseconds, to our frame, embodies such a large Lorentz factor that we see their time frame dramatically stretched.

Cosmic rays, and these secondary particles they produce when high-energy cosmic rays encounter the upper atmosphere, have become the bane of astronomers using equipment on and off the Earth. Most detectors that are highly sensitive to photons – detecting radiation in its "particle" guise – also respond very well to high-energy particles. Exposures with a charge-coupled device pick up numbers of such "cosmic-ray hits" in proportion to the exposure time, making it necessary to break long exposures into shorter pieces to be able to reject them even at the expense of the ultimate depth of the exposure (Fig. 3.10). This is the factor which limits single exposures with the cameras on the Hubble Space Telescope to about 45 minutes, even when observing the small regions of sky that it can view for longer periods. Even though the solar magnetic field diverts low-energy particles before they reach the Earth's vicinity, we see quite enough of these particles. In an early analysis of the effects of cosmic rays on *Hubble* images, colleague Rogier Windhorst noted that the number of such impacts with various levels of energy follows a power law (that is, the number of events scales with a power of the energy, in this case a negative power). In hindsight, that should have been obvious – the fact that synchrotron radiation from our Galaxy shows a power-law spectrum with energy of the radiation means that the electrons, and probably other kinds of particle, follow a power-law distribution. [14] From the radio emission of the Galaxy directly into the silicon chips of our cameras... (Some charge-coupled devices do not wait for cosmic rays. One batch of these chips was made using a glass substrate with an unusually high uranium content, so that no amount of external shielding could reduce the number of energetic particles degrading the images.) At least ground-based telescopes benefit from shielding by the atmosphere, even though this can be a mixed blessing.

These particles can be even more disruptive to detectors that have to be sensitive to smaller amounts of energy, such as need to be used for infrared astronomy. The camera systems on the European Space Agency's *Infrared Space Observatory* (ISO) could have their sensitivity altered for many minutes by a cosmic-ray strike, which occurred frequently enough to produce a dictum for ISO data analysis: "If you haven't thrown away half the data, you can still do better." Observing strategies for ISO, which could be quite complex and involve coordinated motions of the telescope and filter mechanism, were largely dictated by the need to retrieve enough redundant information to reduce the effects of these "upsets". More recent missions (such as *Spitzer*) do better both through improvements in the detectors, and improved ways of reading them out and processing the data streams.

3.5 Synchrotron radiation from the galaxies

The suggestion that we might see synchrotron radiation from cosmic sources dates at least to work by Hannes Alfven and N. Herlofson in 1950. Still, some of the theory had yet to be worked out. In 1953, Iosif Shklovsky[15] in the USSR showed that synchrotron radiation could show uniquely high degrees of polarization.[16] This was shown to be true for the diffuse visible light from the Crab Nebula, by comparing photographs made with a polarizing filter rotated to different angles, to a degree not explained by any other kind of radiation. The role of synchrotron radiation gained appreciation by 1962, as Gart Westerhout and colleagures presented extensive data on the polarization of radio waves from

---

[14] Except that low-energy particles are deflected out away from the inner Solar System by the Sun's magnetic field; their place is taken locally by particles accelerated in the Earth's vicinity.

[15] To whom is attributed the description of theoretical astrophysicists as "often in error, never in doubt".

[16] Considered as a wave propagating through space, radiation can have a preferred direction of "vibration" - that is, of oscillations of the electromagnetic field. Various directions of polarization may be enhanced by some processes for producing the radiation, or its further interaction with matter. The most familiar is probably the polarization produced when light scatters off solid surfaces, which allows polarized sunglasses to reduce visual glare.

Fig. 3.10. Cosmic-ray effects on raw *Hubble* images. Both upper and lower images show the same part of the galaxy cluster Abell 2125, from one of the CCD chips on the telescope's Wide Field Planetary Camera 2. The upper image is from a single 1,300-second exposure, showing the deleterious effects of charged particles passing through the detector. The lower exposure has been processed by combining two such exposures and omitting bright features which appear only in one. This algorithm is equally effective at removing cosmic-ray detections and fast-moving asteroids. (Data from the NASA/ESA Hubble Space Telescope, original principal investigator F.N. Owen.)

the Milky Way. This kind of radiation is produced copiously wherever relativistic particles spiral through cosmic magnetic fields. We see it in supernova remnants, the jets of radio galaxies, quasars, and microquasars, and in the widespread diffuse radio emission from the Milky Way. When they pass within the Solar System, we call the radiating particles "cosmic rays".

Synchrotron radiation is most important at the longest wavelengths that pass unimpeded through the interstellar gas – that is, for radio astronomy. Going to shorter wavelengths, progressively more energetic particles and stronger magnetic fields are required to produce strong synchrotron emission. This does happen in the jets of a few radio galaxies (such as M87, Fig. 2.11), where electrons have Lorentz factors as large as a million, and the continuum of the Crab Nebula.[17] In fact, the Crab Nebula and its pulsar accelerate particles to such high energies that its synchrotron spectrum extends smoothly through the X-ray into the gamma-ray regime. This makes the Crab the strongest steady source of this high-energy radiation and a basic calibration object. It was long the custom for high-energy astronomers to report the intensity of newly discovered objects as so many tenths of a Crab, or so many milliCrabs.

The energy locked up in synchrotron sources of radio galaxies and quasars can be enormous, including the relativistic particles and the magnetic field alike. To be conservative, many calculations assume a minumum-energy condition, which arises when equal energies are tied up in these two components. Folding in the additional energy needed to create the lobes by pushing material into the low-density gas between galaxies, such a cloud can contain 10^{59} ergs, which is the energy that 300 million stars like the Sun would release over their entire 10-billion-year lifetimes. The energy input from active galactic nuclei is vast, and long-lasting. It must persist for at least a few millions years at a time, to move this material outward, and probably longer, since the advancing edges meet with resistance from intergalactic gas.

Synchrotron radiation provides examples at once of both relativistic time transformation and beaming. These are not actually separate issues, both resulting from the transformation between the emitting particles' frame and our observer's reference frame, although in doing the calculations we tend to think of them as conceptually distinct.

Further reading:

The relevant history of discovery in radio astronomy is summarized in Salter and Brown, chapter 1 of *Galactic and Extragalactic Radio Astronomy*, 2nd edition, ed. G.L. Verschuur and K.I. Kellermann, Springer, 1988.
Michael W. Friedlander, *A Thin Cosmic Rain: Particles from Outer Space*, Harvard 2000 (rev. ed.).
Jay Holberg, unpublished manuscript on Sirius and white dwarfs.
Synchrotron radiation lab background: http://www.nsls.bnl.gov/about/history/
Jocelyn Bell Burnell's account of the discovery of pulsars is given in "Petit Four", Eighth Texas Symposium on Relativistic Astrophysics, *Annals of the New York Academy of Science*, 302, 685-689, Dec. 1977, and reproduced in http://www.bigear.org/vol1no1/burnell.htm. This includes her comments on students and supervisors.

---

[17] As one might expect from the foregoing discussion, high polarization of their light is one key sign of synchrotron radiation.

4 From shifting stars to multiple quasars

One of the most sensational predictions from Einstein's general theory of relativity was the gravitational deflection of starlight. This has no counterpart in classical physics. Since the masses of both interacting objects enter into Newton's expression for gravitational attraction, a massless entity (such as light) cannot respond to the force of gravity. In contrast, Einstein showed that gravity can be viewed, not as a force between objects, but as a change in the geometry of the playing field – spacetime itself – in which objects exist. Motion, in this view, always takes place in the straightest available path, which may be curved if the straightest path is not exactly what we would think of as "straight". Light, too, might have to transit in such a path. This penetrates to the central insight of general relativity, which has been described as "Matter tells spacetime how to curve, and the curvature of spacetime tells matter how to move". But it is not only matter that takes instruction.

Introductory physics and astronomy textbooks are littered with drawings of rubber sheets stretched by weights – a simple analogy for the way the curvature of spacetime dictates the paths available to even particles with no mass that could be affected by Newton's kind of gravitational force. The analogy serves well in some ways, but, like many such teaching illustrations, can be subtly misleading if taken too literally. What the rubber sheet gets right is that curvature of space means that there is more space in certain regions than "ought" to fit there if we extrapolate the geometry farther away in emptier areas. It does this by curving in the direction perpendicular to the normal flatness of the sheet – a curvature which attracts our intuition because the weights on the sheet respond to the same gravity we are trying to illustrate and sag downward (Fig. 4.1). (It always seems to be downward – I have yet to see one with the curvature producing a dimple pointed upwards, although that would make exactly the same point). This can lead to the idea that this curvature must extend into a fourth dimension of space – one perpendicular to the three ordinary dimensions. After all, was not Einstein's work mixed up with some kind of fourth dimension? Here we are stretching the analogy (if not the rubber) to the breaking point. There is no physical viewpoint corresponding to the way we see the sheet outside its curved (and ideally two-dimensional) shape. We can only see what amounts to a view vertically down (or up) onto the sheet, inferring curvature only from how objects and radiation pass through various parts of it. (Jeff Hester at Arizona State University has taught this by using a rubber sheet which is viewed only

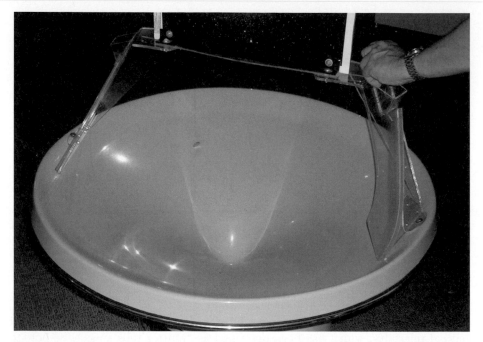

Fig. 4.1. A "wishing well" designed to simulate motion in a gravitational field, with the same kind of curved surface often used to illustrate spacetime near a massive object. In this one, coins spiral inward much like matter near a black hole. Such analogies have both strengths and weaknesses; for example, the curvature described by general relativity does not pass into another spatial dimension.

via an overhead TV camera, so that only the changed motion of things rolling across it can be seen).

4.1 "Lights all askew in the heavens"

This view – that the playing field of spacetime is bumpy where matter has altered it – leads to a prediction which is strikingly different from expectations based on the old idea of gravity as a force acting between objects with mass. Even an entity with no mass at all, moving at the speed of light, must change its direction when passing through a spacetime lump, because the "straight lines" along which light travels cannot be straight in all places without showing a net bend around massive objects in a sense, inheriting this curve from the surroundings.

Under normal conditions, this effect is miniscule even by astronomical standards. The only hope for measuring it that Einstein could predict would be for light from distant stars passing near the surface of the Sun (Fig. 4.2). Light on such a path would be deflected from its "straight" path – that is, the one it would have taken through space were the Sun not there – by the scant angle of 1.75

arcseconds.[1] The phenomenon might be observable during the brief span of a total solar eclipse, when telescopes could photograph distant stars seen through the solar corona. In essence, the experiment is simple. Observe a field of stars with the Sun (and Moon) in front of them, and then six months later in the night sky, look for a small displacement of the stars' locations outward from the Sun seen during the eclipse.

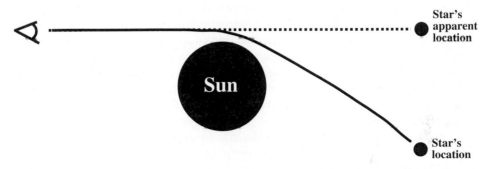

Fig. 4.2. Schematic diagram of the gravitational deflection of starlight observed during a solar eclipse. The solid line traces the path of deflected light, while the dashed line indicates the apparent direction of a star. The effect has been dramatically exaggerated for clarity.

In this case, relativity made a prediction with no fudge factor – since we know the mass of the Sun, and its radius, we have an unambiguous calculation of the expected deflection of the light. It also predicts that the angle of deflection should drop off inversely with distance the starlight passes at its closest to the Sun. This dependence provides an extra bit of the theory to test, as well as a convenient way to combine results from various stars whose light has passed at different distances from the solar surface. Results of eclipse experiments were generally expressed as the equivalent displacement of a fictitious star whose light just grazes the solar surface, based on the actual measurements of all stars measured and their different angular displacements.

Confirmation of Einstein's prediction made front-page headlines in the *New York Times*, reporting "lights all askew in the heavens". Nearly as newsworthy as the results were the dramatic circumstances of the confirmation. The solar eclipse

---

[1] Astronomers continue to use a system of angular measure handed down from ancient Babylonia. A circle is divided into 360 degrees. Each degree is divided into 60 minutes of arc, each of which consists of 60 seconds of arc. The sixties come about because 60, and especially 360, are evenly divisible by more integers than any other numbers of comparable size – a useful feature if you have yet to invent a way to represent and manipulate fractions. As an added bonus, one degree is very nearly the average motion of the Sun against the background stars during one day. The apparent sizes of Sun and Moon are both close to 30 minutes of arc. 1.75 arcseconds is the about how small the finest lunar details resolved by a good telescope appear, or about how large Jupiter's biggest moon Ganymede looks from our vicinity.

of May 29, 1919, was nearly ideal for this purpose. The eclipse was unusually long, reaching a duration of 6 minutes 51 seconds for the crucial total phase, with the Sun standing high in the sky from a path curving from the tropics of South America to central Africa. During totality, the eclipsed Sun stood in front of a rich star field on the outskirts of the large, bright cluster of stars known as the Hyades, providing many more potentially measurable stars than at most eclipses (Fig. 4.3). The best-known observations of this eclipse were headed by Sir Arthur Stanley Eddington of Cambridge University. The potential intellectual import of the results was great enough to send expeditions packing to the Atlantic islands of Sobral, off Brazil, and Principe, near the African coast, to check on a theoretical prediction made by a scientist who was a citizen of what had lately been an enemy nation. A dozen stars near the darkened Sun could be measured on their photographs, with a combined deflection (referenced to our imaginary star whose light just grazed its edge) of 1.8 ± 0.2 arcseconds, agreeing with Einstein's calculations as precisely as the telescopes could measure. [2]

The Cambridge expeditions were the first to measure this deflection, but not the first to plan such measurements. German mathematician and astronomer Erwin Freundlich had long been in communication with Einstein, who noted that he was the first scientist to seriously consider new tests of relativity (the subject of a 1916 book by Freundlich). Backed by the head of the Krupp industrial empire, he planned an expedition to photograph a solar eclipse observable from Feodosia on the southeastern coast of the Crimea. This eclipse occurred on August 21, 1914. But at the end of June, bullets in Sarajevo set off the chain of events that would embroil Europe in war by the first week of August. Already en route, Freundlich found himself interned in Russia. He might have been better off chancing the weather as the eclipse crossed Norway and Sweden. Had he done so, he might have even scooped Einstein, whose earliest predictions are said to have been incompletely derived, amounting to only half the later (and observationally confirmed) value.

The experiment was repeated at eclipses for decades, building up more accurate average values. Eclipse measures using visible light were eventually surpassed by measurements of distant objects with radio telescopes, which need not wait for an eclipse and achieved enough accuracy to measure the deflection of radiation passing quite far from the Sun. This is fortunate, because the plasma in the solar corona can introduce its own deflection of radio waves. While it is possible to allow for this extra deflection if observations are made across a wide frequency range in a short time, the experiment is cleaner if one can simply avoid

---

[2] There is a revisionist tendency to accuse Eddington of something between spin and fraud in his treatment of the data here, since the individual measures were not of the highest quality. However, these accusations tend to come from sources who otherwise have an axe to grind about relativity in general. His data treatment does not appear to have been unusual in the pre-computer era, when shortcuts in the interest of easing calculations were routine tools of the scientific trade. A much more interesting episode involves Eddington's ideas on stellar evolution being taken so seriously as to mask problems in the first measurement of the gravitational redshift, which has been taken up in Chapter 3.

Fig. 4.3. The sky behind the eclipsed Sun, with its corona represented schematically, as seen from the island of Principe on May 29, 1919. Sun and Moon stand against the northern edge of the bright Hyades cluster. The brighter members, as well as the brilliant foreground star Aldebaran, make up the familiar V-shape forming the face of Taurus the Bull, at the bottom of this 10-degree view. Sky map produced using the *Guide8* software, by Bill Grey of Project Pluto (projectpluto.com).

looking through the dense inner part of the corona. By 2004, such radio-source measurements showed the the deflection angle is within 0.04% of the original prediction based on the Sun's mass and size. At the same time, improvements in our ability to measure the positions of stars in visible light led to seeing gravitational deflection all over the sky, rather than only for light passing close to the Sun. The European Space Agency's *HIPPARCOS* satellite, in the course of its three-year mission to measure accurate positions and distances of more than 250,000 stars, measured the deflection of starlight for groups of stars more than 90° from the Sun in our sky. This was one effect that had to be removed to reach the project's goals for accuracy in measuring stellar positions and their changes.

Variations could be rung on the gravitational deflection of starlight. It was easy enough to calculate how much starlight would be deflected upon passing, say, Jupiter, with the result that it was not worth looking for such a tiny effect. As late as 1972, Steven Weinberg's definitive text on gravitation indicated that there was no realistic hope of measuring the 0.002-arcsecond shift in the apparent direction of stars whose light passed just above the Jovian cloudtops. (This tiny angle is about the apparent size of a local cosmic icon – 0.002 arcseconds is

Fig. 4.4. Einstein with the observatory staff at the 40-inch refracting telescope of Yerkes Observatory on May 6, 1921. (Yerkes Observatory, used by permission.)

A scientific couple provides another striking example of how people outside the mainstream of research institutions have contributed to the idea of gravitational lensing. Jeno Barnothy (1904–1996) and Madeleine Barnothy Forro (1904–1993, noted as the first woman to earn a physics Ph.D. in Hungary) emigrated to the USA from Hungary in 1948. Married for 55 years, their joint research interests spanned cosmic-ray physics, effects of magnetic fields on organisms, and astrophysics (including, eventually, gravitational lensing). In work from the early 1960s, they examined the potential impact of gravitational lensing on our view of the Universe. Before the announcement of the first discovered gravitational lens, they had published fifteen papers or conference presentations dealing with gravitational lensing by galaxies. They were particularly interested in whether all quasars could be less luminous objects hugely amplified by the gravitational deflection as their light passed through the most effective regions of closer galaxies, and appearing to vary in brightness as the foreground galaxy's motion lensed different parts of the nucleus most strongly. In the early days of observing gravitationally lensed quasars, everybody who was anybody in the field had a letter from Jeno Barnothy complaining that they had fundamentally misinterpreted the situation. (I count myself slightly proud to have such a letter from 1983 – two, in fact). The 1970s saw a simmering level of discussion on gravitational lensing – not only from people who could list reasons the Barnothys had to be wrong, but from uninvolved astronomers wondering whether pairings of quasars in the sky could represent lensing of one by the other.

4.3 Lensing revealed – the double quasar

As so often happens in science, the first discovery of gravitational lensing was a byproduct of another kind of research entirely. In this case, it was the continuing quest of radio astronomers for perspective on just what they were seeing. Even as radio telescopes could be linked to improve the positional accuracy of their measurements, and reveal structural details of the radio emission, there was a pressing need to find out what kinds of supernova remnants, galaxies, and even more exotic quasars gave off the radio emission. Such "source identification" would bring all the tools of astrophysics to bear on these objects, including distance and chemical measurements possible with visible light but not the radio emission by itself. This was especially desirable as sky surveys reached fainter radio sources, some of which might lie at cosmologically large distances and be able to tell us about such large issues as the curvature of space and the age of the Universe. Each new, deeper radio survey was accompanied by efforts to optically identify the objects responsible – often a slow process, because the radio positions could be uncertain enough to encompass several potentially related optical objects.

The key object was first noticed during a survey of the northern sky, carried out around the end of 1972 by Dennis Walsh. He used the famous 250-foot Lovell radio telescope at Jodrell Bank, in the English countryside south of Manchester. In the first of a string of serendipitous events that Walsh wrote hastened the discovery, director (later Sir) Bernard Lovell was so impressed by the program that he allocated several extra weeks for a northern extension of its sky coverage, passing 56° north of the celestial equator near the nose of the Great Bear. Among numerous undistinguished radio sources, he catalogued one blip on a strip-chart record by its approximate coordinates, and maintained it as part of a list of 800 sources for further study. Walsh's colleague Richard Porcas observed these sources further with the 300-foot radio telescope (Fig. 4.5) of the US National Radio Astronomy Observatory in Green Bank, West Virginia.[3] The next lucky break came here. The source we care about, named 0957+561 after its approximate celestial coordinates in the reference year 1950, is close in the sky to the bright spiral galaxy NGC 3079, which is itself a significant radio source. The two were blended together in the original observations; by itself, 0957+561 is not strong enough to have been recognized in those first sky scans. Porcas

---

[3] This telescope, built on a shoestring in 1958, was intended to provide only an interim capability to observe objects as they passed due north or south until more capable instruments were completed. It lasted until its sudden collapse brought on by metal fatigue in a bearing, 30 years later. It has recently been replaced with the 100×110-meter, fully steerable, Robert C. Byrd Telescope, named in honor of the Senator who assured its funding.

pointed the 300-foot telescope, at a higher frequency and thus with sharper resolution, at the indicated position and found... nothing. The original radio source turns out to have been a blended combination of lensed quasar images, to the north, and the nearby active galaxy NGC 3079 to the south (see Fig. 4.6). The next piece of fortune in the tale is that Porcas used a search strategy to find "missing" sources by first moving the telescope to the north – finding the quasar images, 0957+561, before going south (at which he would have probably concluded that the radio source was NGC 3079 recorded with a larger-than-usual position error).

Fig. 4.5. Banks played a key role in the discovery of the double quasar 0957+561. At right is the 250-foot Lovell telescope at Jodrell Bank, south of Manchester in the UK. Across the Atlantic was the 300-foot radio telescope of the US National Radio Astronomy Observatory at Green Bank, West Virginia, sited by design nowhere near a large city.

In both the UK and USA, research students (Anne Cohen and Meg Urry, respectively, who became co-authors of the same journal publication) worked at noting potential counterparts of these radio sources on the photographs of the Palomar Sky Survey, which remained for forty years the best available references for the appearance and content of any desired area of the sky. Made almost simultaneously in the summer of 1976, their notebooks show the first evidence of something unusual associated with this radio source. Nearby, they found a pair of blue starlike objects separated by about 6 arcseconds. Such blue images were almost always quasars, placing this source high on a list for followup with the optical spectra that would reveal the nature and redshifts of the images. The Jodrell Bank radio astronomers had working with colleagues worldwide in seeking such spectra, involving telescopes at Palomar, McDonald Observatory in west Texas, and eventually Kitt Peak in Arizona. Walsh and Cambridge coworker Bob Carswell wrote a successful proposal to use the 2.1-meter telescope (then usually called "the 84-inch"; Fig. 4.7) for this project, with a system that could display the accumulating spectral data in real time. The northern image came up during their first clear night on 29 March, 1979, and within 20 minutes it was apparent that this is a quasar with redshift near $z = 1.4$.

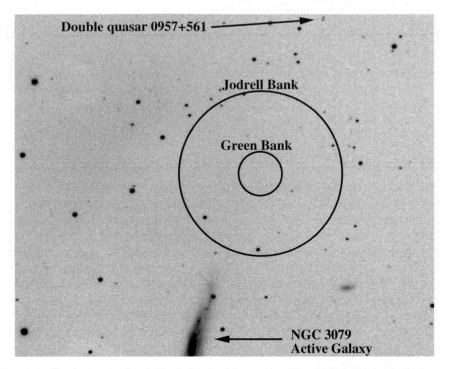

Fig. 4.6. Confusion in the field of the double quasar. The underlying optical picture, shown as a negative for clarity, shows the double quasar Q0957+561 near the north edge, and the active galaxy NGC 3079 about 0°.3 to its south. The circles indicate the 50% response radius of the systems used for observations at Jodrell Bank and Green Bank. The original Jodrell Bank survey blended radio emission from the two sources and gave an intermediate position, roughly as shown here, requiring a more extensive search from Green Bank to find the sources. (Background image: University of Alabama 0.4m telescope.)

Since the southern image had appeared nearly as blue on the Sky Survey pictures, Walsh and Carswell decided also to observe it. When its spectrum appeared virtually identical, they first suspected a problem or miscommunication in setting up the telescope. As telescope operator Barbara Schaefer recalls, "The night that we made the original measurements was notable in that the seeing was quite a bit better than normal. In those days standard seeing at Kitt Peak was usually about 2 or 3 arcseconds. That night it was about 1 arcsecond or maybe a little bit better. This was important as the separation between the two objects was not very large and we were worried about light from one spilling into the aperture while we were integrating on the other. I felt confident that with seeing that good we could separate one from the other. Bob and Dennis weren't totally convinced when the two showed up with the same spectra! We were excited when the first one showed as a QSO. When the second one did also they started looking for what could have been wrong as not only was it also a QSO, but the spectra looked essentially identical. They explained to me how

Fig. 4.7. The 2.1-meter (84-inch) telescope of Kitt Peak National Observatory, in southern Arizona, used by Walsh and Carswell in the discovery of the double quasar 0957+561. Moonlight highlights the dome during working hours.

unlikely and unusual that was. We did go back and forth a few times from one to the other before they were convinced that we were not getting contamination between the two."

By the next night, conversations with University of Arizona astronomer Ray Weymann had brought the gravitational-lensing idea to the fore. Memories differ as to whether it was at midday or night lunch. In those days, night lunch at Kitt Peak was an institution in itself, where astronomers could sneak a break if collaborators kept the telescope going on clear nights, or gather for conversation and information on cloudy nights. Cooks would be on duty to provide a snack to order until well after midnight, and the exchanges between astronomers from astronomers using 10 telescopes from as many institutions were an important way to exchange ideas and the latest results in those pre-Net days. Weymann was using the University of Arizona's 2.3-meter (90-inch) telescope on the same mountaintop starting the following night. with a setup configured to examine a narrower slice of the spectrum with greater wavelength resolution. Even as Walsh and Carswell returned to the pair each clear night to build up better data, Weymann found that both images showed the same pattern of narrow absorption features due to intergalactic hydrogen in front of them. Carswell notes, "Whether or not we expected absorption I can't remember either, but it was there in both, and at identical redshifts and the same apparent strength. So we thought we had a lens, with Ray's data playing the most compelling part in making us think that was the probable explanation."

Ray Weymann had been aware of gravitational lensing for a decade. In fact, he had set graduate student Richard Wilcox on the trail in the late 1960s, noting that, as fanatical as Jeno Barnothy in particular could be on the topic, there was nothing wrong with his physical understanding or lensing equations, and that it would be scientifically worthwhile to do a detailed calculation of what the statistics of gravitational lensing ought to be, based on the numbers of quasars and galaxies known. Wilcox had made some progress on this problem when he was found dead, at his own hand, in early 1969.

After Carswell and Walsh left the mountaintop, observatory director Geoff Burbidge noted that their proposal for radio-source identifications had been one of several quite similar proposals received. Given the vagaries of allocation committees, this may be one more link in the chain of luck that brought us the unveiling of gravitational lensing in 1979 rather than 1989 or 1999.

This interpretation raced through the astronomical community, with a buzz of interest among astronomers at Kitt Peak diffusing quickly through the community. Pam Walsh remembers an unusual excitement in Dennis's voice as he phoned home (still a tedious and expensive proposition in 1979). I first heard of it as a graduate student working at Lick Observatory, where a postdoctoral astronomer recently arrived from Tucson described the situation of identical quasar spectra with a rather profane variation in the Cockney "infix" usage to intensify the meaning [4]. The suggestion that we were actually seeing twin images of one quasar suggested additional tests of the lens picture, as well as more observations that would become possible if this were confirmed. Most obvious, if we see images, we should also see the lens – a galaxy, or perhaps group of galaxies, in the right place to deflect the light from a single background object in two directions that we see coming to us from 6 arcseconds apart on the sky.

The two quasar images differed by only about 25% in brightness. Intuitively, that suggested a lens galaxy which would be seen about midway between the two images. Nothing obvious showed up on ordinary photographs, but then again, a galaxy at the best distance to split the quasar light might well be 50–100 times fainter than each quasar image. Astronomers in Hawaii and California pursued disparate ways to find such a lens galaxy.

Alan Stockton at the University of Hawaii could use the growing suite of telescopes atop 4,200-meter (13,800-foot) Mauna Kea, whose exceptional quality as a site for astronomical observation has only become more obvious in ensuing years. He had devoted considerable effort to using image intensifiers (the distant predecessors of night-vision systems) to augment traditional photography, making the most powerful and precise use of the university's 2.2-meter (88-inch) telescope (itself the first sizeable instrument on Mauna Kea). His system for digitally scanning and calibrating the photographic plates let him do some of the kinds of image analysis that became routine for most of us only with the widespread arrival of solid-state imaging devices several years later. When he

---

[4] Along the lines of I-bloody-dentical.

use this arrangement to photograph the double quasar, Stockton recalls, "I remember seeing the fuzzy object next to QSO image B when I first turned on the light after developing the first plate, and being somewhat puzzled, since I had naively assumed, from the equality of the lensed images, that the galaxy would be roughly halfway between the two. I called up Ray Weymann to get his take on what I had found, and he suggested that I talk to Peter Young. I called Young up, and I was somewhat crestfallen when he said something like: 'Yes, that is a galaxy with redshift 0.39...' "

Peter Young was leading a group at Caltech, taking advantage of their larger telescope at Palomar and early access to digital imagers to pursue ways of seeing the lens galaxy that did not require the exquisite image quality that was usual from Mauna Kea. Their suspicions had been aroused by subtle differences between the spectra of the two quasar images – differences which could be beautifully explained if their spectrum of the southern image was contaminated by light from an elliptical galaxy at a redshift near $z = 0.4$. Long-exposure images, obtained using one of the first charge-coupled devices (CCDs) available to astronomers, revealed a faint halo of light around this image, centered only 0.7 arcseconds to its north, and matching what we would see of such a galaxy if its inner bright parts were lost in the glare of the quasar light. Their images also showed additional, very faint, galaxies; the lens galaxy was at the center of a substantial cluster, all near the same redshift of $z = 0.39$. These galaxies began to show up in the best pre-CCD images (Fig. 4.8), and have now been studied in some detail.

Peter Young was remarkable for his genius, even in a profession noted for attracting intellectual heavyweights. He began graduate school at the University of Texas at the age of 21, where Gerard de Vaucouleurs recalled him as "one of the brightest students I ever knew". He transferred to Caltech the following year (after writing or coauthoring four research papers, including his first essays into the growth of black holes). He finished the Ph.D. program in a mere three years, and after another year as a postdoctoral researcher, was hired as a Caltech faculty member in 1979 (at the unheard-of age of 25). His style of research and presentation, combining detailed analysis of digital measurements, numerical modelling of the object under study, and deep theoretical interpretation, showed how the rest of us would be striving to write papers five years later. Young made major contributions to our understanding not only of gravitational lensing, but to the presence and effects of massive black holes in galaxies, the evolution of gas over cosmic time, and the physics of cataclysmic binary stars. These contributions are all the more remarkable in view of his death by suicide in late 1981, at the age of 27. We will see more of his legacy in Chapters 5 and 7.

Returning to the situation in 1979, the data had given – there was a perfectly plausible lens galaxy - and the data had taken away – the lens galaxy was not remotely halfway between two nearly equal images. How could light passing such different environments be amplified almost equally? Calculations tailored for light passing through a galaxy, rather than outside its distribution of stars, quickly showed how intuition based on small and opaque lenses (that is, individual stars) could lead us astray. For a galaxy, whose mass is spread over an immense volume through which light can pass, there is a certain "miss distance" for which the combination of mass to one side and its distance give the greatest bending angle. Passing either closer or farther than this will give smaller angles – in one case because the light passes through a less curved region of space farther from the mass, in the other case because mass to one side partially offsets the mass to the other. This means there can be a path going quite close to the center of the galaxy which is only slightly bent (as indeed happens in 0957+561). There can even be a third image very close to the galaxy center, but this one is demagnified rather than magnified, making it very dim and seldom seen. The magnification is related to the change in bending angle with "miss distance" of light rays; strong changes in this quantity mark areas of high magnification as well as strong amplification of the object's intensity,

Uniquely, gravitational lensing is *achromatic* – all wavelengths of light and all kinds of radiation are affected equally. This follows from all kinds of electromagnetic radiation having the same speed in a vacuum (to which intergalactic space is a very good approximation) and experiencing the same curved spacetime as they traverse it.[5] This fact set off attempts to detect the images at every possible wavelength. Since they were first detected as radio sources, they were clearly strong enough for detailed study by radio telescopes. In particular, the National Radio Astronomy Observatory had recently completed the 27-antenna Very Large Array (VLA) on the plains of San Agustin in New Mexico (Fig. 4.10). This array was capable of imaging pieces of the radio sky in sharper detail than astronomers were accustomed to seeing with optical telescopes, understandably furnishing a unique facility for radio astronomers worldwide. A group led by Bernard Burke from MIT, one of the driving forces behind the concept of the VLA, turned the new array on the double quasar, and found a puzzle. The two quasar images appeared as brilliant points of radio emission, but their surroundings differed. The northern image showed twin lobes of emitting material spanning several arcseconds on each side, typical of many quasars, which were missing from the southern image. Burke in particular was rather skeptical of the lensing hypothesis until this could be shown to be a natural situation. Once again, the peculiar behavior of light passing through a galaxy (instead of around a star) caused the problem (or came to the rescue). Knowing the location of the lensing galaxy and something about its mass, from the Mauna Kea and Palomar measurements, one could show that a double image would be produced only of objects in a small region behind it. While the quasar itself is inside the magic region (or else we would not have recognized it as interesting), most of the large radio lobes fall outside this; they are distorted in our view, but not

---

[5] Indulging my professor's license to be pedantic, I will add that this is true only for radiation with wavelengths much shorter than the size of the lensing object. For radio waves with the unimaginably long wavelength of 100,000 light-years and longer, the lensing effect is reduced because most of the wavelength would lie far from the center of curvature of this spacetime dimple. Since no waves longer than about a kilometer in wavelength penetrate the interstellar gas in our galaxy or the solar wind in our own Solar System, I am happy to ignore this complication.

Fig. 4.8. Our improving views of the double quasar, before the era of solid-state imagers and space observatories. The top photograph was taken in 1903 with Lick Observatory's 0.9-meter Crossley reflector, while the middle picture used the same telescope in 1979. The bottom image used an image intensifier at the Kitt Peak 2.1-meter telescope. This last picture just shows light from the main lensing galaxy around the lower (southern) quasar image, and some of the surrounding galaxies as well. Compare these images to the *Hubble* image in Fig. 4.9.

Fig. 4.9. The gravitationally-lensed quasar Q0957+561 in a color-composite *Hubble* image. The quasar images, 6 arcseconds apart in our sky, appear blue, with the redder light of the primary lensing galaxy almost in front of the southern image. Additional galaxies are part of a surrounding cluster, at redshift $z = 0.39$ or a distance of about 4 billion light-years. Data from the NASA/ESA Hubble Space Telescope archive, originally obtained with G. Rhee as principal investigator. This figure appears as color plate 4.

doubled. Burke pursued further work on this and other lens candidates, so energetically that the meeting convened to honor his 60th birthday was entitled simply "Gravitational Lenses".

Ray Weymann led a more detailed study of the visible-light spectra, using the substantially greater light grasp of the then new Multiple Mirror Telescope, or MMT, on Mount Hopkins south of Tucson. This unique telescope was designed by astronomers at the Harvard–Smithsonian Center for Astrophysics and the University of Arizona, using a unique military-surplus opportunity to gain many of the capabilities of a large telescope at a fraction of the cost of traditional designs. Cancellation of the US Air Force Manned Orbiting Laboratory (MOL) program, designed as a piloted reconnaissance satellite and essentially made obsolete by advances in automated systems, freed up the six 1.8-meter (72-inch) mirrors that would have been used in the MOLs' downward-looking telescopes. The MMT incorporated all six in optically separate telescopes on a single mounting, bringing the images together for a narrow field of good definition with a brightness equivalent to that from a single 4.5-meter mirror. This setup was ideal for measuring spectra of stars and quasars, where the limited

Fig. 4.10. The antennae of the north arm of the Very Large Array stretch into the distance, distorted by a summer mirage on the high plains of New Mexico. The whole instrument includes 27 dishes, which can be deployed in a Y-shaped pattern spanning as much as 36 km in extent. Since its completion in 1978, the VLA has been a centerpiece of radio astronomy not only for US investigators, but for astronomers worldwide.

field was not a limitation.[6] The MMT results confirmed the virtual identity of the two images, down to minute differences between the mean shifts of different spectral features that matched between the two, and the identical patterns of absorption from gas in front of the quasar (although not all that far in front). They wrote that these data strengthened the lens interpretation, particularly their finding that the redshifts of foreground absorbing clouds were identical to within 15 km/s (one part in 20,000) – difficult to arrange if we see the lines of sight tens of thousands of light-years apart but much easier, and almost inevitable, if the lines of sight never separate as far as they seem to from our point of view. This point is reinforced by the more detailed ultraviolet *Hubble* spectra shown in Fig. 4.11.

One more test of the lensing hypothesis had been theoretically shown to tie in with another fundamental issue on astronomy: the scale of the Universe. The light-travel times of the various images in a gravitational lens will differ by a small amount (something like one light-year out of billions). Resfdal had shown in the 1960s that, if we can understand the lens properties precisely enough, the time delay between the two images is proportional to the lens distance, and thus the scale of the Universe. [7]

---

[6] The MMT has recently been refitted to use a single 6.5-meter primary mirror on the same structure and mount, although the acronym was retained for continuity. When constructed, the "classic" MMT was the world's third-largest optical telescope; with the flurry of telescope construction driven by new optical and mechanical possibilities beginning around 1990, the new MMT is currently tied for place number 13, but has the oldest mechanical parts of the telescopes in the top 15!

[7] See Chapter 8 for more on Einstein, Hubble, and the expanding Universe.

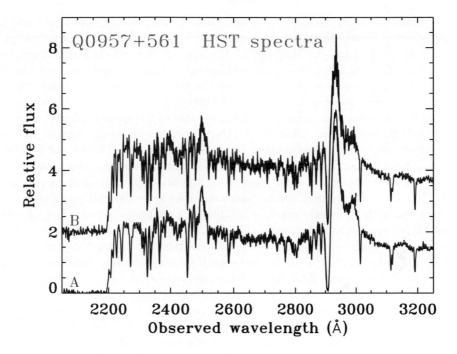

Fig. 4.11. The spectra of the two lensed components of the double quasar Q0957+561, obtained by *Hubble* in the ultraviolet range. These show the identity of the two images' spectra, matching not only the quasar emission peaks from hydrogen (Lyman α, which we observe redshifted to 2918 Å) and oxygen (the far-UV line of O^{5+}, observed at 2498 Å), but numerous narrow absorption features from gas in foreground galaxy halos and the intergalactic medium. These plots combine seven spectra each from *Hubble's* Faint-Object Spectrograph with one each from the newer Space Telescope Imaging Spectrograph, which are plotted individually although overlapping. The spectrum of image B has been offset upward by 2 units for clarity. The cutoff in light from each image shortward of about 2190 Åresults from hydrogen in a cloud slightly in front of the quasar; subtle differences in the accompanying hydrogen absorption lines are the only features in the spectra which do not match exactly, as the two lines of sight pierce slightly different parts of the foreground cloud. (Date from the NASA/ESA Hubble Space Telescope archive, originally obtained by Robert Bless and Ross Cohen.)

Since most quasars flicker in brightness on times of weeks and longer, it could be profitable to follow the two components of 0957+561 for such variations. At the very least, seeing the same pattern of variability in both images, at different times, would be virtually irrefutable evidence that a single background object was being lensed – and if Nature cooperated, could offer a way to jump the whole traditional, and somewhat rickety, ladder of steps leading to the Hubble constant and the scale of the Universe. I played a small role at this point, using a series of photographs extending over about 15 months, taken with Lick Observatory's

0.9-meter (36-inch) Crossley reflector. These could be augmented for historical perspective by additional photographic plates taken with the same telescope as early as 1902, thanks to NGC 3079 – the same galaxy which had first confused the identification of the radio source. Photographs using the Crossley's full field of view centered on the galaxy included the quasar images (albeit distorted, as seen in the top panel of Fig. 4.8), near the edge of the plates.[8] The old plates sampled the long-term average brightness ratio of the quasar images, while the monthly pace of the new data showed that the southern image became relatively brighter and then gradually faded over a span of months. While the ideal goal would have been to continue watching to see whether the same pattern could be seen on both images, I did not want to let this study run on much longer, being aware that the group at Palomar was also looking for variability, using a larger telescope with the vastly superior solid-state technology represented by one of the first charge-coupled devices (CCDs) in regular use for astronomy. These devices are much more sensitive and precise than photographic emulsion. The first generation, many of which were spinoff prototypes from initial development for what became the Hubble Space Telescope, were starting to find their way into major observatories around 1980. So I published what evidence I had, at least demonstrating the variations in the two images and setting a minimum time delay that was allowed. Multiple programs around the globe have continued to monitor the brightness of the twin quasar component for twenty years now, establishing the time delay (about 1.2 years, just slightly longer than my old limit from 1980) and showing evidence for what kinds of stars and other bodies are present in the lens galaxy.

The whole suite of lensing signs was now evident for the double quasar 0957+561 – identical spectra, a lensing galaxy, lensed structure at least from the radio to the ultraviolet in the regions appropriate for the galaxy's gravity, and (eventually) correlated variability. This finally led to the payoff on a side bet between Dennis Walsh and another quasar identifier, Derek Wills at the University of Texas. On looking at survey pictures of the potential radio source counterparts, Walsh had written on Wills' blackboard, "No QSOs, I pay Derek 25 cents. One QSO, he pays me 25 cents. Two QSOs, he pays me a dollar". Derek and Beverley Wills obtained improved spectra with the McDonald Observatory 2.7-meter (107-inch) telescope, scarcely the worse for wear after having holes shot in its primary mirror in 1970. Plots of their spectra arrived up in Dennis Walsh's mailbox with a silver dollar taped to them, but Derek declined to accept his 75 cents in change even though, after all, there is only one quasar.

Further candidate lenses among quasars followed quickly. Since quasar cores are so small that even an elongated and distorted one looks like a point of light, the symptom of lensing continued to be multiple quasar images with identical

---

[8] A 1915 photograph of NGC 3079 formed part of the survey in which H.D. Curtis demonstrated not only the ubiquity of "spiral nebulae", but the patterns that showed the existence of absorbing dust in their flat disks even before we could show that these are in fact independent galaxies. Oddly enough, observers throughout this early period failed to record their identities in the logbooks; Curtis's identity as the observer was established by Tony Misch based on handwriting comparison.

spectra. Ray Weymann and coworkers turned up another within a year – not just double, but triple, and on closer inspection quadruple. He recalls, "we were observing PG 1115+080 at the 90-inch and the seeing suddenly got very good, and we were astonished to see the single object resolve into three separate objects. We tried to get spectra, but the object was low in the sky by that time and the seeing got bad again. As soon as we could we got on the MMT with good seeing and we got decent spectra of all three objects. I remember Roger Angel was also up at the telescope and nicknamed this triple Mickey Mouse because it looked like a nose and two eyes." Later observations showed that the "nose" consists of two quasar images a fraction of an arcsecond apart. The memorable designation has not stuck, but the importance of gravitational lensing certainly has.

Discoveries of first lensed quasars, and then lensed galaxies in which we could trace the distortions in their image shapes, have followed so rapidly that routine use is made of the phenomenon, not as a test of General Relativity, but as an astronomical tool. We can now use, if not quite God's own telescope, certainly Fritz Zwicky's. Its gaze has told us about dark matter from our own galaxy to distant clusters, and about cosmic history on the grandest scale.

Further reading:

For Mandl's life and role, Jüergen Renn and Tilman Sauer, "Eclipses of the Stars – Mandl, Einstein, and the Early History of Gravitational Lensing", 2000, preprint 160 of the Max-Planck-Institut fuer Wissenschaftsgeschichte (to appear in *Revisiting the Foundations of Relativistic Physics: Festschrift in Honour of John Stachel* (Kluwer).
Einstein's 1936 paper is "Lens-Like Action of a Star by the Deviation of Light in the Gravitational Field", *Science* **84**, 506
Freundlich's expedition to the Crimea is discussed in Amir C. Aczel, *God's Equation* (Delta Books, New York), 1999.
Zwicky's note on lensing by galaxies is "Nebulae as Gravitational Lenses", *Physical Review* **51**, 290 (1937)
Dennis Walsh, "0957+561: The Unpublished Story", in *Gravitational Lenses*, Springer-Verlag Lecture Notes in Physics 330, 1989, p. 11.

Colour Tables

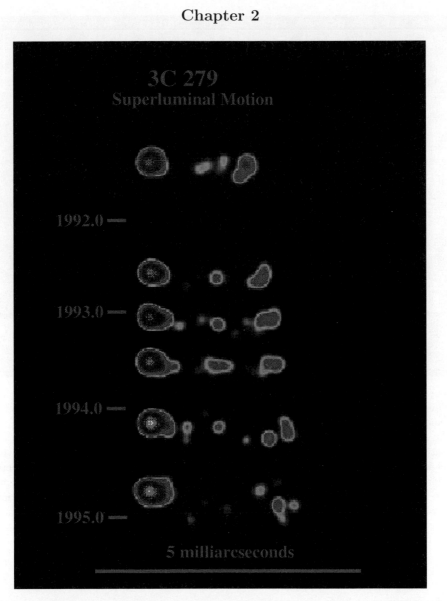

Fig. 2.13. The quasar 3C 279, showing superluminal motion in its jet. Pseudocolor intensity coding has been used to make subtle features more apparent. The prominent outer knots to the right are moving outward with a projected speed of $4c$. At this quasar's distance of 5 billion light-years, the scale bar of 5 milliarcseconds corresponds to a length of 100 light-years. The bright knot to the right moves almost 20 light-years in 4.5 years, as viewed in our reference frame. These observations, at a radio wavelength of 1.3 cm, were carried out initially with an *ad hoc* network of radio telescope, and starting in 1994, with the Very Long Baseline Array of the National Radio Astronomy Observatory, a dedicated network of telescopes stretching from the Virgin Islands to Hawaii. (Images courtesy Ann Wehrle and Steve Unwin.) This figure also appears in color as Plate 1.

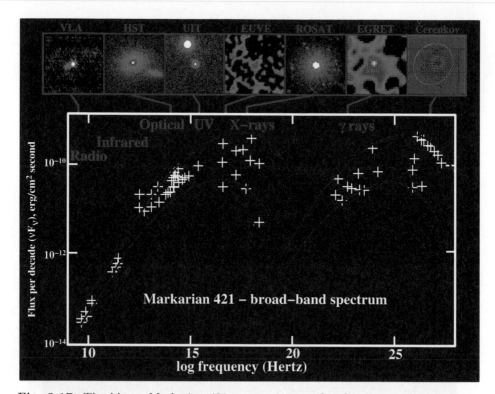

Fig. 2.17. The blazar Markarian 421 as seen across the electromagnetic spectrum. This montage shows the relative energy received from this blazar in various spectral regions, both when it is quiet and when it is in a bright flare such as may accompany the appearance of new jet material. The inset images show its appearance to various instruments used for these observations. The distinct gamma-ray peak is evidence that its lower-energy radiation is being scattered and Doppler-boosted in a jet directed nearly toward us. (Data provided by T. Weekes, the NRAO FIRST radio survey, and the NASA archives from the *Hubble, Compton, ROSAT, UIT,* and *Extreme Ultraviolet Explorer* missions.) This figure appears in color as Plate 2.

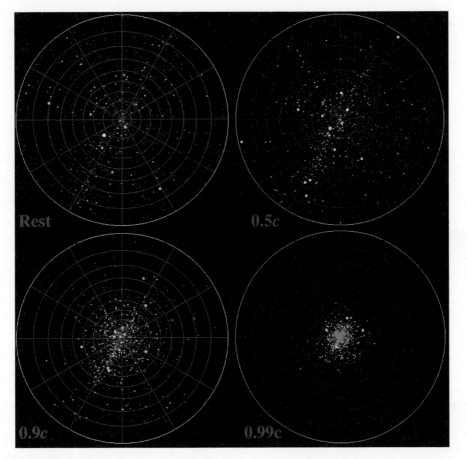

Fig. 2.27. Simulations of the starbow as seen looking forward from the bridge of a starship in the vicinity of Earth, heading toward Orion. Each shows a full hemisphere's view. One image is at rest (that is, our actual view), while the others represent forward velocities of 0.5, 0.9, and 0.99c. As the velocity toward Orion increases, aberration bunches the stars forward at high velocities, and Doppler shifting changes their colors and relative brightness, depending on the broad shape of each star's spectrum. At yet higher speeds, infrared emission from interstellar dust shifts into the visible band straight ahead. (Images courtesy of Jun Fukue of Osaka Kyoiku University, originally produced for the 1991 Japanese-language book *Visual Relativity 2*.) This figure appears as color plate 3.

Chapter 4

Fig. 4.9. The gravitationally-lensed quasar Q0957+561 in a color-composite *Hubble* image. The quasar images, 6 arcseconds apart in our sky, appear blue, with the redder light of the primary lensing galaxy almost in front of the southern image. Additional galaxies are part of a surrounding cluster, at redshift $z = 0.39$ or a distance of about 4 billion light-years. Data from the NASA/ESA Hubble Space Telescope archive, originally obtained with G. Rhee as principal investigator. This figure appears as color plate 4.

Chapter 5

Fig. 5.8. Simulated images of a galaxy (Messier 83, in this case) as it would appear gravitationally lensed by a point mass (such as an intervening black hole). These three images depict the situation as the lens moves from left to right across the galaxy, eventually forming a ring image of its bright core. This and Fig. 5.9 use screen shots created with Peter J. Kernan's "Lens an Astrophysicist" applet. (For clarity, I have removed superfluous images inside the Einstein ring, created by the applet's assumption of periodic tiling of the sky with the same pattern outside the image region.) This figure appears in color as Plate 5.

Fig. 5.9. Simulated images of the galaxy Messier 83 as it would appear lensed by foreground point masses, now varying the lens mass between images. The lens mass doubles for each step downward in the figure. As in Fig. 5.8, I have removed superfluous images inside the Einstein ring. This figure appears in color as Plate 6.

Fig. 5.12. The spectacular lensed galaxy in the cluster Cl 0024+1654, in a color composite *Hubble* image. The cluster generates five images of a single blue background galaxy, distinguished by its distinct "theta" shape; in this case, the lensing effect is visually obvious. (W. Colley, Princeton University, and NASA). This figure appears in color as Plate 7.

Chapter 8

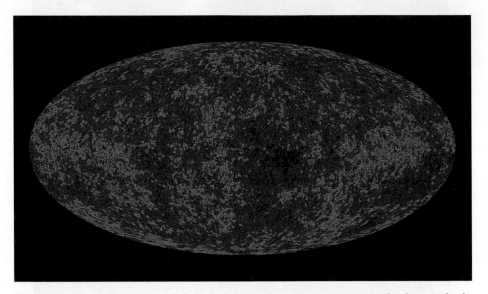

Fig. 8.13. An all-sky map of fluctuations in the cosmic microwave background, the product of the first year of measurements by the *Wilkinson Microwave Anisotropy Probe* (WMAP). Pseudocolor mapping is used to make these subtle patterns stand out more clearly. This map has been processed to remove the signatures of foreground dust and gas in the Milky Way as well as more distant radio sources. (NASA/WMAP Science Team.) This figure appears in color as Plate 8.

5 Through the gravitational telescope

With the case secure that gravitationally lensed quasars exist, astronomers turned to further questions. How many quasars are lensed? Does lensing alter our view of the most distant objects? Are the lensing objects mostly individual galaxies, groups, or clusters, and what does lensing tell us about their masses? And finally, can we actually use gravitational lensing to extend the range of our telescopes, as Zwicky proposed almost seven decades ago? The answers have been fruitful enough to make gravitational lensing a lively research area, and a tool in regular use for the study for cosmology and galaxy evolution.

As lensing makes it easier for us to study the distant amplified objects, it also gives us pause as to how precisely we can measure anything in the distant Universe. In what seems like the revenge of the Barnothys, the brightest quasars at great distances will preferentially be those which have been amplified by lensing, making it a delicate proposition to reconstruct their "typical" properties. In the presence of gravitational lensing, we do not see a perfect image of anything in the high-redshift Universe; there are lumps of matter everywhere, so at least some slight image distortion must always occur. We can at best look for evidence as to when this distortion is important, and find that in fact it can be very informative. The signature of this *weak lensing* allows us to trace the large-scale distribution of matter in the Universe, whether dark or bright,

5.1 Quasars double, triple, quadruple, and sextuple

An obvious path, once the first few lensed quasars had been found, was to see how many would turn up in systematic, controlled surveys. The incidence of such lenses would tell how many galaxy-mass objects exist in the Universe, whether they contain visible stars or not. "Dark lenses" would be a fundamental discovery – collections of matter which had failed to form visible galaxies or clusters. And the redshift distribution of the lensed quasars would tell, in essence, how much space there is in various redshift ranges, and therefore what the expansion history of the Universe has been. Roughly, individual galaxies would split images by about an arcsecond - not always easy to resolve from the ground. Galaxy groups and clusters could split images by 5–20 arcseconds; the double quasar 0957+561 with an image separation of 6 arcseconds has the effect of its main lensing galaxy substantially augmented by the surrounding cluster.

One group at the University of Arizona surveyed bright quasars with the most straightforward of techniques. On a night when the atmosphere allowed calm and sharp images, they inspected the quasars using a TV camera, noting any that showed structure. Having found that one of the six mirrors of the Multiple-Mirror Telescope produced noticeably better images than its fellows, they did some of the survey with five of them closed off, adding to the advantages of the site and telescope enclosure's construction in providing unusually high-quality images.

A more wholesale approach was taken by Bernard Burke's radio-astronomy group at MIT, eventually led by Jacqueline Hewitt. They started with a low-resolution survey for radio sources, done with the Green Bank 300-foot telescope, which included over 20,000 sources, and later extended to the southern sky using the Parkes 64-meter dish in Australia. They then used the giant VLA interferometer for quick "snapshot" observations of 8,000 sources which might be quasars. The image quality of these individual snapshots was not very high, but it did not need to be to recognize the characteristic multiple points of a lensed quasar. By now, these surveys have turned up an even dozen lensed quasars – doubles and triples, some with measurements of the time delay between images as we see the quasar's own flickering first in one image, then another whose light path is slightly longer.

A strong point of the radio surveys is that, although only a small fraction of quasars has strong radio emission, the ones that do could be found through the absorbing dust of a lens galaxy, removing a potential bias from selection based on optical light alone. Indeed, some lensed images are bright in the near-infrared but sufficiently reddened by dust to almost vanish in optical images.[1]

The most complete optical search for gravitational lenses has been one of the many products of the Sloan Digital Sky Survey (SDSS). The SDSS was the brainchild of astronomical jack-of-all-trades James Gunn, now at Princeton. At a meeting on sky surveys hosted by the National Optical Astronomy Observatories, he realized that technology (particularly in his tinkering hands, as a veteran of breakthrough instrumentation for Palomar and *Hubble*) had reached a point allowing a deep, digital survey of most of the sky in just a few years. This survey would reach ten times deeper than the earlier landmark Palomar Observatory Sky Survey and its southern-hemisphere extensions, with all the data in calibrated digital form, and encompassing images in five rather than two filters. The scientific payoff could be enormous. The data could be mined for asteroid surveys, faint nearby stars, the large-scale distribution of galaxies and its correlation with galaxy properties, and the evolution of quasars. As the data archive is populated, this puts us one step closer to astronomers (and everyone else) having the sky on a desktop.

---

[1] There have been studies comparing the colors of pairs of lensed images at various redshifts to study whether the character of dust in galaxies changes with redshift. The idea is that, for a genuine lens, a difference in color implies that one is seen through more dust in the lensing galaxy, and thus the difference tells us the relative amount of absorption at various wavelengths, which in turn tells us something about the mix of dust grains in the galaxy. If the data were not already in hand for other reasons, that would definitely qualify as doing it the hard way.

As it developed and was finally funded by the Alfred P. Sloan Foundation, the SDSS included a spectroscopic survey of roughly a million galaxies and 100,000 quasars, found from their unique color signatures in the survey's images. These have already included the highest-redshift quasars yet known. A dedicated wide-field telescope of 2.5-meter aperture was built at Apache Point, New Mexico; the camera and spectrograph alternate in use, depending on atmospheric conditions. The survey is still in progress, but has already turned up statistically complete sets of lenses in some parts of the sky. The most interesting may be SDSS J1004+4112, a quadruply imaged quasar at $z = 1.7$ with images as much as 14 arcseconds apart. This must be an example in which an entire cluster of galaxies is responsible for the lensing. Indeed, a cluster of galaxies at $z = 0.68$ is found in front of the quasar, a cluster which the lensing alone shows must be dominated by unseen dark matter.

These same searches turned up some *faux* lenses - pairs of quasars with similar but not quite identical redshifts, important spectral differences, or grossly dissimilar X-ray or radio properties. Lack of any plausible lensing galaxies in very deep observations can also suggest that an image pair is not, in fact, a gravitational lens (although the possibility of a "dark lens" must still be considered). These must then be genuine binary quasars, which have their own astrophysical interest (although it took George Djorgovski at Caltech a long time to convince some of the rest of us that we should care). Genuine binary quasars are interesting because they give hints as to the evolution and triggering of the whole quasar phenomenon. In the nearby Universe, quasars are extremely rare in galactic nuclei. But as we wind the cosmic clock back, the density of quasars increases, peaking around redshift $z = 2$, 10 billion years ago. At that epoch, there were enough quasars that we may find several in a single cluster of galaxies. But having both members of a galaxy pair host quasars at once stretches random statistics; it requires that both be set off at once. The trigger event may well be the gravitational interaction of two galaxies. Abundant evidence in the nearby and distant Universe shows that galaxy collisions can disturb a galaxy's gas enough to drive up the rate of starbirth, lighting up the galaxy in a brilliant burst of star formation. There have tantalizing hints that something similar happens for quasars, radio galaxies, and Seyfert galaxies (Chapter 7), albeit taking much more statistical finesse to interpret. Perhaps the strongest such hint came from early *Hubble* imaging of the galaxies around quasars. A study led by John Bahcall of the Institute for Advanced Study found that over one third of the quasars' galaxies not only had close companions, but a very particular kind of close companion – compact galaxies very close to the quasar galaxy, sometimes appearing within its structure. These kinds of galaxies (somewhat like M32, the close companion to the Andromeda galaxy) can survive tidal disruption and penetrate very deeply into another galaxy, causing strong tidal forces within a small part of it. Binary quasars may be systems in which both nuclear black holes begin rapid accretion and erupt into activity within a short time: a hundred million years.

Examining quasars for more sets of lensed images, too close together to distinguish using ground-based telescopes, was something that the Hubble Space

Telescope could do well even in its original aberrated condition. Such a search formed Hubble's first "spare-time" project, inaugurating what came to be known as its snapshot capability. Once the scheduling software had packed in all the observations from a primary list that it could fit into a slice of time, the constraints of solar array direction, orientation, and time to move to the next target would leave frequent, but brief, gaps. These gaps, usually a few minutes long, would allow motions across the sky of a few degrees and quick exposures with one of the telescope's cameras, as long as it did not have to take the additional time required to lock onto a pair of guide stars. Short-exposure images of targets in accessible parts of the sky could be chosen from a long enough list, provided that their science goals could be reached with the relaxed stability provided by the telescope's gyroscopes alone without the exquisite precision of its star-tracking sensors. Even with images significantly trailed in one direction by residual errors in the gyroscopes, and aberrated by the misfigured primary mirror, as simple a pattern as a pair of starlike images could be found unless the direction of image trailing happened to wipe out the pairing. A group led by John Bahcall of the Institute for Advanced Study used these snapshot opportunities to examine quasars from a list of several hundred; a basic feature of snapshot projects throughout *Hubble's* mission has been that you have no control over which ones of your objects might actually be observed, since they are chosen to fill holes in a predetermined schedule.

The snapshot team learned interesting things about *Hubble* operations as well as the population of gravitationally lensed quasars. Some of the early observations completely missed the target quasar in the narrow field of the Wide Field/Planetary Camera. In a case of relativistic irony, the problem was traced to neglect of the effects of aberration from Hubble's orbital motion in the initial gyroscopic pointing. This was not an issue for ordinary observations, since the effect was taken into account in handing over tracking to the precise Fine-Guidance Sensors, and having it corrected in the snapshot mode opened up a new way to use these snippets of time (a mode which has been employed for thousands of bright targets over the observatory's lifetime).

Of the 498 quasars of which images were obtained, only two proved to be split into multiple gravitational images (Fig. 5.1). Another nine potential faint companion images turned out to be chance alignments with foreground stars. *Hubble* has delivered excellent and crucial data on many lensed quasars, but all were discovered in other ways – ways which allowed surveys of thousands rather than hundreds of quasars.

The award for "most complicated gravitational lensing of a quasar" currently goes to the sextuply-imaged object B1359+154, where we see six images as a $z = 3.2$ quasar shines through galaxies (Fig. 5.2). A galaxy group near $z = 1$, of which several are apparent in the *Hubble* image, does most of the work. Modelling suggests that there are three more demagnified images yet to be found.

Careful examination of lensed quasars, especially in the infrared where the glare of the quasar light is less of a problem, has turned up distorted images of the host galaxies around the quasars, opening an avenue for studying them

Fig. 5.1. One of the few gravitationally lensed quasars turned up in the *Hubble* snapshot survey. This image pair has a separation well under 1 arcsecond (the length of the scale bar), as seen in this guided followup image. The lensed quasar is very distant, redshift $z = 3.8$ corresponding to a lookback time of about 12 billion years. (J. Bahcall/NASA.)

using the lens magnification. Some of the host galaxies are lensed into nearly complete Einstein rings, when we can look to faint enough levels.

5.2 Through the lens of serendipity

We have learned to be on guard for the role of gravitational lensing in any extremely luminous object in the distant Universe. This has made several noteworthy discoveries less fantastic than they first seemed, since gravitational lensing meant that these objects were not as powerful as initially calculated. Many of these most spectacular lenses were found serendipitously, in the course of normal astronomical surveys.

One prime example is the infrared source known as IRAS FSC 10214+4724. The designation indicates one of the million or so objects seen in the sky survey performed by the US–UK–Dutch *Infrared Astronomical Satellite* (IRAS), which carried a cryogenically cooled telescope into orbit in 1983 to survey the sky at wavelengths inaccessible from Earth. Using hitherto classified detector techniques, IRAS brought far-infrared astronomy from being the province of gifted cryogenic engineers to something the entire astronomical community approached on an equal footing. The IRAS source catalogs were publicly released, in one of the first examples of a shift in astronomical sociology. Everyone had equal access to these results, and after the satellite's helium supply was exhausted, no one

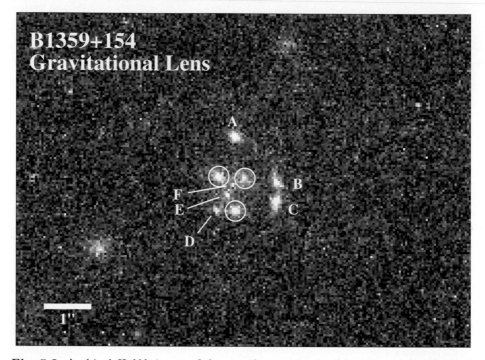

Fig. 5.2. Archival *Hubble* image of the complex gravitational lens system B1359+154. Six images of a distant quasar at $z = 3.2$ are seen (A–F), with three galaxies (circled) near $z = 1$ as the main contributors to the lensing effect. Some of the quasar images show arc segments of the Einstein ring, indicating that the light from its host galaxy is also being amplified. This image was taken in deep red light, to improve sensitivity to normal starlight from galaxies at high redshift. (Data from the NASA/ESA Hubble Space Telescope archive; original imagery obtained in 2000 with Chris Impey as principal investigator.)

could improve on the data. This introduced a way of working to which we have become accustomed after using the archives from *Hubble, Chandra, ROSAT*, and other space missions, and now adopted by some ground-based observatories as well. It is well to remember what a step forward it is to be able to find and distribute data digitally, compared to the days when the only original record of an observation might be a single photographic plate, catalogued or not, languishing in someone's desk drawer.

To a degree far beyond what its planners dared suggest, IRAS did revolutionize astronomy. Its well-calibrated sky survey allowed its use for cosmology – galaxies it detected formed a fair sample of the Universe far less subject to biases due to dust absorption than any previous selection. A sample of galaxies selected from the IRAS data would give a fairer sampling in both depth and position on the sky than any previous technique. On this basis, a UK group, led by Michael Rowan-Robinson of Queen Mary College in London (and a member of the original IRAS science team), undertook an ambitious project (the

Fig. 5.3. The *Infrared Astronomical satellite* (IRAS), a US–UK–Netherlands mission which surveyed the far-infrared sky for eleven months in 1983–4. The instrument, seen pointing leftward behind an angled sunshade, was a 60-cm telescope cooled by a liquid helium tank. In a near-polar orbit, IRAS was constantly oriented pointing outward, with the solar arrays (top) pointed at the Sun. This geometry swept its field of view around the entire sky every six months, building up a nearly complete survey at wavelengths of 12–100 μm. This engineering duplicate is in the National Air and Space Museum in Washington.

"QDOT survey") to measure the redshifts of 2,163 galaxies from the IRAS data, randomly selected from the entire data set (they picked one in six above a fixed infrared brightness). While most of these galaxies are bright enough that their redshifts could be measured with, say, 1.5-meter telescopes and the technology of the time, the total time to complete such a survey would be prohibitive (which is to say, they would still be working on it). This group realized something that had previously appeared the secret of astronomers doing a few specialized kinds of stellar astronomy. If your telescope points rapidly and accurately, it can be very efficient to use a much larger telescope than "usual" to examine bright objects, allowing observations of dozens per night in quick succession. The problem then becomes not so much obtaining all the data, but the bookkeeping.

The team used the then new 4.2-meter William Herschel Telescope (WHT; Fig. 5.4) erected on La Palma, in the Canary Islands. This site enjoys excellent atmospheric conditions, which has made it the *de facto* site of a "European Northern Observatory". The first telescopes there were part of a UK–Netherlands collaboration, relocating the 2.5-meter (98-inch) Isaac Newton Telescope from Sussex and adding the 1-meter Jacobus Kapteyn Telescope. The 4.2-meter WHT went into service in 1987, employing the latest in computer control, including an altazimuth mount which offered great cost savings at the expense of more elaborate and constant mechanical motions.[2] In 1991, on the final night of a 40-night project at the WHT, the team found a galaxy at redshift $z = 2.3$, making it by far the most distant object discovered from the *IRAS* data (a distinction it still holds) and, including the infrared component, the most luminous object found up to that time. This extraordinary energy output, and the fact that so much of its radiation was absorbed by dust grains and re-emitted in the infrared, made it a strong candidate for a protogalaxy.

Fig. 5.4. The 4.2-meter William Herschel Telescope (WHT), perched near the rim of the Caldera de Taburiente on the island of La Palma. This location delivers unusually steady imaging, which has continued to make this site attractive for new facilities. The WHT incorporates an altazimuth mount, providing for a smaller and less expensive mounting and building at the expense of more complex computer control for tracking. Until the era of 8-meter telescopes, this was the flagship optical telescope for British and Dutch astronomers.

The astronomers' search for newborn galaxies in the early Universe – protogalaxies – can be fairly compared to the quest for the Holy Grail. We are still hunting, with the prize being understanding of when galaxies came together, when they started making stars, and how long an initial epoch of rapid and violent star formation persisted. Recent results bring us ever closer to this period, but it is not certain that we are yet seeing far enough back in time to see most

---

[2] The WHT also marked the telescopic swan song of Grubb Parsons, a firm which had been responsible for some of the most technologically advanced telescopes for a generation.

of the action. For IRAS-F 10214, the infrared luminosity suggested an enormous rate of star formation, thousands of times what the Milky Way sustains today. Further observations with millimeter radio telescopes showed correspondingly strong emission from the molecular gas which fuels starbirth – another hint that this was a truly extraordinary object, perhaps of a kind found only for a brief time early in cosmic history.

This object attracted wide attention, and new data suggested that all might not be as it seemed. By 1995, infrared images from one of the 10-meter Keck telescopes showed a suspicious configuration. Two of the several objects seen in earlier images took on the form of a partial bullseye, with a faint object near the center of an arc stretching perhaps 30° around it. This would be expected if some still very exceptional, but not unprecedented, background galaxy had its light amplified and distorted on passing through the gravitational influence of a foreground galaxy (itself perhaps several billion light-years distant). If this were a lens, all the estimates of its luminosity, star-forming rate, and gas content had been inflated by some (perhaps very large) factor. Soon thereafter, a *Hubble* image (Fig. 5.5) removed virtually all doubt, showing a very clear arc shape forming perhaps 1/4 of a complete Einstein ring. Modelling of the situation suggests that we see this galaxy brightened by perhaps 30 times due to lensing. It is still a very bright and young galaxy, and important for study because it does appear so bright – but not nearly as powerful as we first thought.

Perhaps the most spectacular lensed quasar was found in a study of the redshifts of nearby galaxies. A team at the Harvard–Smithsonian Center for Astrophysics (CfA) spent several years measuring the redshifts of all bright galaxies in various slices of the sky, using these data to map the galaxy distribution. These data furnished the conclusive evidence for vast voids, sheets, and filaments of galaxies, in a first glimpse of the frothy texture of the Universe. In 1984, during an observing session with the 1.5-meter Tillinghast telescope which CfA operates in Arizona, a galaxy near the southern border of Pegasus showed a spectrum utterly different from the normal run of galaxies. The resident observer showed it to team member John Huchra, who happened to be using the Multiple-Mirror Telescope at the top of the mountain. Indeed the spectrum should not have looked familiar – it was that of a quasar at redshift $z = 1.7$, coming from the middle of a large, nearby galaxy. Announcing the discovery at a January 1985 meeting in Tucson, Huchra said that his first reaction had been the feeling that "maybe Arp was right", referring to Halton Arp's long-time contention that quasars are not at the great distances suggested by their redshifts, but are relatively nearby objects which can be ejected from the nuclei of galaxies. At that point, only a single quasar image could be seen on the ordinary pictures available. The team did consider models with a single highly-amplified gravitational image near a galaxy nucleus, but these proved to be unnecessary. Telescopes at the best sites soon revealed the configuration which came to be known as the "Einstein Cross" – four quasar images arranged around the nucleus of the galaxy itself. This is spectacularly resolved in *Hubble* images (Fig. 5.6), called by one Web visitor "the eeriest astronomical image I've ever seen".

Fig. 5.5. A *Hubble* image of the high-redshift, lensed infrared galaxy IRAS F-10214+4724. The background galaxy lies very close to the central point for gravitational lensing by the foreground galaxies, especially the elliptical galaxy at the center of the arc's curvature, and is accordingly distorted into a portion of the Einstein ring. This distortion is correspondingly accompanied by an amplification of the background galaxy's total brightness, making it unusually amenable to detailed study. Data from the NASA/ESA Hubble Space Telescope archive; original principal investigator B.T. Soifer.

The very fact that such distinct spectral features from the quasar showed up in the original spectrum showed that the spectra must be quite similar. Indeed, observations taken from Mauna Kea and La Palma under the calmest skies (Fig. 5.7) show spectra that are identical except for signatures of light passing through the foreground galaxy – small amounts of reddening by dust, and absorption lines from interstellar gas at the galaxy's redshift. Had the center of this galaxy not been more transparent than some other spirals, we would have been hard-pressed to find the Einstein Cross. The background quasar is "actually" located within 1 arcsecond of the galaxy nucleus; statistics tell us that we are very unlikely

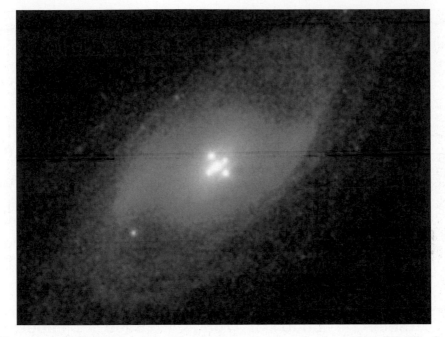

Fig. 5.6. A *Hubble* image of the Einstein Cross quasar and galaxy 2237+030. The four quasar images are cleanly separated from the galaxy core, while some of the bar and spiral structure of the lens galaxy appear. (Data from the NASA/ESA Hubble Space Telescope archive; original investigator J. Westphal.)

to find another alignment so exact involving such a nearby galaxy and bright quasar.

5.3 Lenses, time, and distance

The Einstein Cross has been the subject of careful monitoring to watch for variability of the components and derive the delays between light-travel times for the various images. In this case, such a study is complicated by the light passing so deep among the galaxy's stars that small fluctuations can be introduced as individual stars move near the light paths; we end up learning more about stellar motions in the galaxy than about the lensing geometry in this case, and must continue observations for many years to separate these rapid changes, uncorrelated among the images, from the correlated changes in brightness due to the quasar itself. Sjur Refsdal had pointed out in the 1960s that gravitational lenses offer a way to cut through some of the most contentious issues in astronomy, namely the ladder of distance estimators needed to measure the expansion rate of the Universe (the Hubble constant; Chapter 8). *If* we see a lens that we can model mathematically – well-known image positions, redshifts for both lens and quasar, lens galaxy smooth and symmetric with a measurable mass – the dif-

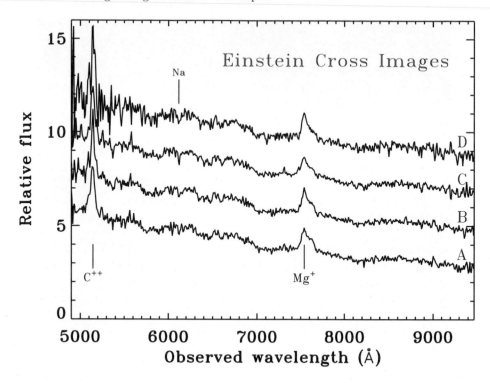

Fig. 5.7. Spectra of the four components of the Einstein Cross gravitational lens Q2237+030, obtained under unusually steady atmospheric conditions with the 4.2-meter William Herschel Telescope on the island of La Palma. The designations A–D are in decreasing order of brightness, as in Fig. 5.6. Small differences in color due to dust in the foreground galaxy have been taken out to avoid overlap of the spectra. Sodium absorption in the lens galaxy is at the wavelength marked with Na. The ripple in the Mg^+ emission line near 7610 Å is an artifact of imperfect removal of an absorption feature produced by oxygen molecules in the Earth's atmosphere.

ference in flight time for light in different images depends only on the distance to the quasar. Thus we would be able to measure a cosmologically interesting distance in a very direct way. The appeal of this technique lies not solely in its simplicity, but in its jumping across a whole ladder of overlapping methods used in the more "traditional" approach to the cosmic distance scale. Any error in one rung of the ladder propagates to all the more distant ones.

Time delays have been measured for several quasar lenses, variously using the quasars' variations in visible light or radio emission. These range from 10 to 400 days, with the original double quasar 0957+561 having the largest value yet measured. After some hard and delicate work to measure indicators of the mass of the lens galaxies from the internal motions of their stars, values of the Hubble constant could be derived. Most of these results gave a smaller Hubble constant (that is, greater distance for a given redshift) than more traditional methods, leading to suspicions in each direction about unknown systematic errors. But we may not need to invoke unknown problems in either analysis. The lumpiness of the Universe introduces lensing effects that are not accounted for in these kinds of model, and would be extremely difficult to evaluate separately. These come about from the many galaxies and clusters of galaxies close to the line of sight, which will collectively produce slight curvatures of spacetime that could add in any direction. One simulation indicates that individual lens systems could have their time delays altered by as much as 50% by foreground and background galaxies, so that a significant ensemble of lenses will be needed to put this technique on an equal footing with more traditional methods. Still, it is reassuring that the distance values from the available time delays are even close to estimates from the redshifts alone.

As convenient as the approximation is for writing the equations of stellar motion, galaxies are not smooth lumps of matter. At the very least, they contain stars. If we look closely enough, these stars make gravitational lensing "grainy", in a sense first shown by Peter Young. Many stars would normally lie within the beam of light reaching us from a lensed quasar, which means that under unrealistically high magnification, its image would show bright cusps and dimmer patches, the total of the small deflections of light around individual stars. As the stars continue in their individual motions through the lens galaxy, this pattern would change, introducing slight variations in the brightness of the whole quasar image we see. The amplitude and rate of such variations would tell us about the population and motions of stars in these galaxies. We have to understand these effects if lenses are to fulfill their promise of measuring distances, since this kind of light variation could either mask or complicate the time delay between major quasar images.

5.4 Dots, arcs, and rings

The geometry of a gravitational lens tells us a great deal about the lensing mass. For quasars, we deal with the number of images and their locations. Since quasar emitting regions are orders of magnitude too small for even *Hubble* imaging to show any extent, we cannot see that their images are distorted into tiny arcs, while for large galaxies we can see the arcs, sometimes blending into complete Einstein rings. Similarly, for lensed radio sources, the small cores are multiply imaged, while the large radio lobes are drawn into arcs and, occasionally, complete rings.

Einstein's]indexEinstein, Albert original calculation shows us the answer for a perfectly symmetric situation: when a source is exactly behind the center of a lensing object whose mass distribution is spherically symmetric,[3] we will see a complete Einstein ring. No direction around the source is different from any other. Departures from this perfect alignment and symmetry will change the ring in interesting ways.

---

[3] More precisely, cylindrically symmetric about our line of sight.

If the lens is off-center to our undisturbed line of sight, the Einstein ring becomes uneven, with two opposite and distinct bright regions. When the background source is small enough, these become two individual images, as in a double quasar. The situation for a point-mass lens (that is, a star or black hole, where all its dimensions are small compared to our resolution) is shown in Figs. 5.8 and 5.9, comparing their lensing effects at different locations in front of the background source and for different lens masses. If the lens is alone and symmetric, these images will appear on exactly opposite sides; displacement from this axis requires "shear", some asymmetry in the mass distribution. This can come from additional nearby galaxies or an entire associated group or cluster.

When the lens is not circularly symmetric, there can be more than two bright points in a distorted Einstein ring. A spiral galaxy with an inclined disk, a group with several massive galaxies, or an elongated galaxy cluster will usually produce more than two images. A group of galaxies can produce very complex image patterns as the rays are deflected in various directions along the way (Fig. 5.2).

If the background source appears too far from the lens, only a single image will be formed, deflected as in Einstein's original discussion of starlight near the Sun. This image will also be slightly distorted and amplified. But when the rays pass far from the lens, these small effects may be detectable only statistically, for large ensembles of objects.

For galaxies without an extremely massive central black hole, pure topology guarantees that there will be an odd number of images. This was shown in a particularly elegant way by the late Bill Burke, a mathematician thriving in an astronomy department at the University of California in Santa Cruz. As in a glass lens, the paths of the rays we see are constrained to leave the background source and reach us. Combining that with the deflection angle for any mass distribution gives a curve of deflection with miss distance, and we may see images where this curve intersects a line along which the rays will actually reach our location. He submitted this work, in his trademark elegant and spare style, to the *Astrophysical Journal Letters*, a journal which values brevity. In this case, however, the editors asked him to lengthen his paper so as to fill most of one page (which the revision did – exactly).

So why do we see so many double and quadruple lenses, with odd-numbered cases very much in the minority? The topology tells us where they may be, but the lensing details remain to determine the brightness of each image. For the great majority of triple-image sets, two will be highly magnified and amplified in brightness, while the third will come from rays passing near the galaxy center and only slightly deflected. Such an image will be demagnified, appearing much fainter than the background object would in the absence of the lens. Teasing out such images can require careful modelling of the lens galaxy to find a faint pinpoint shining almost through its core.

The more extended the background object, the easier it is for some part of it to be at the magic location for a complete Einstein ring. With so many large double sources around quasars and radio galaxies, radio observations were the first to turn up nearly complete examples. More recently, optical and near-

Fig. 5.8. Simulated images of a galaxy (Messier 83, in this case) as it would appear gravitationally lensed by a point mass (such as an intervening black hole). These three images depict the situation as the lens moves from left to right across the galaxy, eventually forming a ring image of its bright core. This and Fig. 5.9 use screen shots created with Peter J. Kernan's "Lens an Astrophysicist" applet. (For clarity, I have removed superfluous images inside the Einstein ring, created by the applet's assumption of periodic tiling of the sky with the same pattern outside the image region.) This figure appears in color as Plate 5.

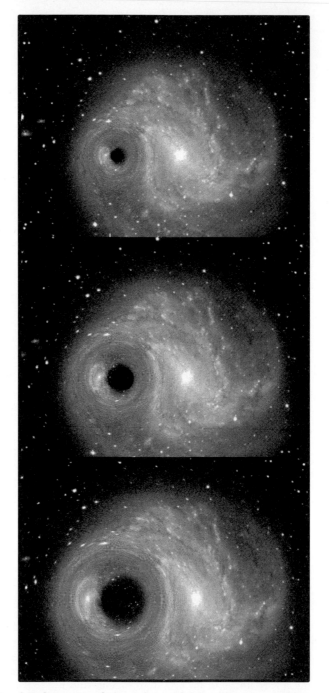

Fig. 5.9. Simulated images of the galaxy Messier 83 as it would appear lensed by foreground point masses, now varying the lens mass between images. The lens mass doubles for each step downward in the figure. As in Fig. 5.8, I have removed superfluous images inside the Einstein ring. This figure appears in color as Plate 6.

infrared observations have turned up equally complete rings of lensed starlight from faint, distant galaxies lying almost exactly behind foreground galaxies.

5.5 Galaxies like grains of... putty

Clusters of galaxies play a special role in our study of the Universe, being very massive, easily seen to great distances, and furnishing the crucible for much of the history of galaxies. Because of their enormous masses, and the fact that we can see through most of a cluster's area, they make very powerful gravitational telescopes, magnifying and amplifying the images of distant objects in ways that we can use to tell us about both the background systems and the clusters themselves. The first became clear with the discovery of blue arcs in several clusters, although it took over a decade for our technology to be able to confirm that we were seeing gravitational lensing.

The first observations of these cluster arcs happened accidentally, in the course of performance tests for newly formulated Kodak photographic plates. Art Hoag, then Director of Kitt Peak National Observatory, had a particular interest in pushing the efficiency of photography for the faintest objects, which he employed in several programs searching for quasars. For another decade, photography would remain the only technique we could use to survey wide areas of sky, until solid-state detectors such as CCDs were available in larger sizes. In October 1975, Hoag exposed a series of new red-sensitive, fine-grain IIIa-F plates at the Kitt Peak 4-meter telescope. His target was the cluster of galaxies Abell 370, which was rich and well positioned for observation, and had previous observations available for comparison. For his purposes, Abell 370 was also interesting in that Jim Westphal from Caltech had recently used it as a sample target for images with the new TV-based Vidicon technology, which Hoag would not have minded showing up. His photographs showed a most unlikely feature – a bright arc, roughly centered on one of the brightest galaxies near the cluster center. Prints from this photograph circulated for years among astronomers, with little progress forthcoming because the arc was so faint that photographic techniques could reveal no more about its nature. With one example before our eyes, astronomers started finding more possibly similar cases, leading to discussion about the nature of these 'cluster arcs'. Multicolor data showed that the arcs are much bluer than typical cluster galaxies, whose visible light comes from populations of old stars (hence one often heard the term "blue arcs").

As more images accumulated, all sorts of theoretical possibilities were bandied about. There were published calculations suggesting that these might be tidally distorted colliding galaxies, made blue by formation of a new generation of stars, or that the arcs were parts of enormous shells blown off by some kind of unimaginably vast explosion. A few people even suggested that these might be gravitationally-lensed, distorted images of background galaxies, seen as their light paths were twisted by the enormous mass of a cluster of galaxies. Seeing a galaxy, rather than a quasar, as the background object meant that we would be seeing the distortion of the image itself – something that happens on a scale still too small for us to see for quasars. This idea made a clear prediction – the spectra of these arcs should show a substantially larger redshift than the clusters, since a lensed object has to be far behind the cluster in order for there to be a long enough light path to see much distortion.

Measuring the spectra of such faint images remained beyond the state of the art for several years. It took the advent of low-noise CCD detectors incorporated into the most efficient spectrographs to make this measurement possible. The answer came almost simultaneously from groups using two telescopes in different hemispheres.

Both groups used very similar instruments and techniques. Roger Lynds and Vahe Petrosian had become acquainted with cluster arcs in 1976, during observations of rich galaxy clusters for tests of cosmology. Their images showed hints of the bullseye pattern that has since become familiar in the cluster Abell 2218 – a pattern which drew gasps when Lynds projected a slide at a 1986 meeting of the American Astronomical Society. However, much of their data had been obtained during engineering trials of new instruments, under conditions "without benefit of what would normally be considered sound observing practice". They returned to study of these arcs in 1985, during trials of a new device designed to improve the quality of CCD images by scanning the camera slowly across the sky during an exposure. These images provided precise enough positions and shapes for the arcs to obtain their spectra, using additional developments in technology, specifically a system in which Lynds had played a key role (Fig. 5.10). Lynds and Petrosian used the 3.8-meter Mayall telescope at Kitt Peak, with the Cryogenic Camera spectrometer. This device (still in service in upgraded form, 25 years later) is built around the CCD chip which collects and stores the light. For maximum sensitivity, astronomical CCDs are usually cooled with liquid nitrogen. What distinguishes this instrument is that, for most effective optical design, the optical elements focussing the light onto this chip are packaged with it in a small vacuum tank which doubles as a Schmidt-type camera. This made possible a higher efficiency than any preceding spectrograph – a crucial consideration when observing faint galaxies right at the edge of our instrumental capabilities. It is painful to contemplate how many photons, carrying priceless information on the distant Universe, survive passage through intergalactic space, clouds of interstellar dust, and the final millisecond's plunge through our turbulent and dirty atmosphere, only to be absorbed by the mirror surface, deflected around the edge of the spectrograph slit, or reflected from metallic coatings on the very detector that is supposed to capture them.

The setup of the Cryogenic Camera allowed astronomers to make custom slits for the spectrograph, which were physically cut into small metal plates for insertion into the telescope. This made it easy to observe many stars or galaxies at one time, by lining up the telescope carefully on reference stars. The same trick could be used to create curved slits to follow oddly shaped targets, maximizing the amount of light actually collected.

Meanwhile, a French group led by Geneviève Soucail had been undertaking their own studies of the arc phenomenon, using the spectacular image quality possible with the 3.6-meter Canada–France–Hawaii Telescope (CFHT)

Fig. 5.10. Roger Lynds adjusts experimental instrumentation in the Cassegrain instrument cage of the 4-meter Mayall telescope of Kitt Peak National Observatory. Hoag's initial photographs revealing lensed arcs in clusters were taken with this telescope, which was later used by Lynds and collaborator Vahe Petrosian to measure the redshifts of arcs and confirm their lensed nature. (NOAO/AURA/NSF, used by permission of NOAO and R. Lynds.)

on Mauna Kea (occasionally seen in the background of Hawaiian snow-skiing posters, as well as in Fig. 5.11). They then attempted to measure the spectra of arcs using the similarly-sized telescope of the European Southern Observatory in Chile, using a spectrograph called EFOSC.[4] EFOSC was similar in spirit to the Kitt Peak Cryogenic Camera, having been started late enough for a staff member of the European Southern Observatory to go to Kitt Peak and ask various astronomers what they would do differently next time. The result was an instrument that allowed more precise subtraction of the spectral features from airglow in the Earth's atmosphere – features that become progressively more important as we look farther into the red spectral region (as we usually do for objects at greater redshifts). As with the Cryogenic Camera, EFOSC could accommodate custom-made slits, including curved examples designed to include all the light of a curved arc. Using 6 hours worth of exposure during the austral summer of late 1987, they obtained a clear detection of spectral features from a much higher redshift than the cluster, submitting this result as strong confirmation of the lensing picture in a manuscript of April 1987. Within a few months, Lynds and Petrosian came to the same conclusion, from very similar results.

---

[4] ESO Faint Object Spectrograph and Camera; the pattern has now proliferated so that various observatories have DFOSC, ALFOSC, HFOSC, and so on.

Fig. 5.11. Geneviève Soucail's team used two 3.6-meter telescopes half a world away from France to reveal the lensed nature of arcs in rich clusters of galaxies. At left is the Canada–France–Hawaii Telescope (CFHT) atop Mauna Kea, renowned for its superb image quality. At right, the otherwise unnamed 3.6-meter telescope of the European Southern Observatory on Cerro La Silla, Chile, which delivered the crucial spectroscopic confirmation of the high redshfts of the arcs. The small auxiliary dome houses a 1.5-meter telescope which can feed light into the main dome's coudé spectrograph when it is not being used by the 3.6-meter telescope.

Bob Goodrich and Joe Miller at Lick Observatory were also chasing the arcs' spectra, using a traditional straight slit set tangent to an arc. Their approach was inspired by the unusually high near-ultraviolet performance of a new spectrograph, which let them look for spectral features where that the other groups were not looking. Unfortunately, the redshifts of their arcs put all the strong emission lines outside the wavelength range they were looking at. Chance favors the prepared, and evidently the wide spectral range as well.

These lensed arcs are nearly one-dimensional stretches of the images of very distant galaxies, so they are very thin. Ground-based images from most telescopes do not resolve their actual thickness, contending with the blurring from our unsteady atmosphere and losing the dimmer arcs into the glow of the night sky. Images from such superb sites as Mauna Kea showed more and more, and finally the refurbished Hubble Space Telescope can see hundreds of arcs around massive clusters. We can now tell which arcs are actually images of the same distant galaxies, when a distinctive pattern shows up with lensing distortions in different directions. The most striking example of this shows up in looking through the very rich cluster Cl 0024+1654. If the cluster were magically removed, we would see a very faint galaxy shaped like the Greek Θ (theta) just about where the cluster center actually appears to us. As we see it through the cluster, this galaxy's light is diverted along at least five different paths to us, each stretched in a different direction. The unusual shape makes it easy to pick out these images (Fig. 5.12). More subtly, some arcs consist of two back-to-back mirror images of a single galaxy, with opposite sequences of knots and color changes. Each of these arcs is a piece of the theoretical Einstein ring, incomplete and departing from being a perfect circle because the cluster's mass distribution is not a smooth sphere and the background galaxy does not lie exactly behind the cluster center. Hubble images can show so many arcs that the eye can trace the shape of the gravitational bending of spacetime, as in Abell 2218 and Abell 1689 (Fig. 5.13). In Abell 2218, we can see the effect of the massive halo of a single bright galaxy, as two lensed arcs part around it (Fig. 5.14). This cluster also amplifies some of the most distant known galaxies, made visible only because of its effect; one object appears to be at a redshift $z = 7$, which we see as it was when the Universe was less than 6% of its present age.

Once the relation between their redshifts and colors (as evaluated across as broad a range of spectrum as possible) has been established, we can use these arcs to reconstruct the distribution of mass in clusters of galaxies. This is especially important since we have found that much of the mass in galaxies and clusters is in some invisible "dark" form (section 5.7). First using ground-based images, and then with *Hubble* data, successively more refined calculations have been able to show us where the mass lies. Perhaps surprisingly, the galaxies are a good tracer, as the same fractions of ordinary and dark matter occur throughout the clusters.

5.6 Realizing Zwicky's dream

By their nature, the arcs we see toward clusters of galaxies trace strong lensing. This means that the total brightness of a distant galaxy is amplified simply because its image appears larger with the same surface brightness (so the amplification factor is nearly the same as the amount by which its length has been stretched into the arc image). Using cluster masses as gravitational telescope, we are finally realizing Zwicky's proposal of extending the reach of our steel and glass by using cosmic masses as part of the telescope. Deliberate searches toward rich clusters have been used to find emission-line galaxies, and sources of radio, submillimeter, and infrared radiation that are gravitationally amplified and would therefore normally be too faint for us to measure. To the extent that we understand the distribution of mass in a galaxy cluster, we can even work out which regions would yield the highest amplification of distant objects and target these areas for the deepest examination.

Use of the "gravitational telescope" has already had some notable successes. One has yielded the most detailed spectrum we have of a high-redshift "normal" galaxy – an object which rejoices in the designation MS1512-cB58. This was originally found in an ordinary survey for members of the galaxy cluster MS1512+36, by a group headed by Howard Yee (and part of the Canadian

Fig. 5.12. The spectacular lensed galaxy in the cluster Cl 0024+1654, in a color composite *Hubble* image. The cluster generates five images of a single blue background galaxy, distinguished by its distinct "theta" shape; in this case, the lensing effect is visually obvious. (W. Colley, Princeton University, and NASA). This figure appears in color as Plate 7.

Network for Observational Cosmology or CNOC, using the 3.6-meter Canada–France–Hawaii Telescope). On closer examination, its spectrum showed it to be at a much higher redshift ($z = 2.7$), so that we see its ultraviolet spectral range arriving here as visible light. The *Hubble* image in Fig. 5.15 shows the slightly bowed arc shape that its image is stretched into; the many cluster galaxies combine to wiggle the simple section of a ring, so that its lensed nature was not immediately obvious. Inspection of the other side of the foreground cluster turns up a counterimage, as well as one of an interacting companion to cB58. Thanks to the lensing amplification, this galaxy is extraordinarily bright for such a large redshift, so that we can measure a much more detailed spectrum than for any other galaxy known at such distances. Two of the world's largest telescopes have been used to obtain uniquely detailed spectra of cB58, allowing the researchers to untangle the stars and gas in the galaxy and its foreground in unmatched

Fig. 5.13. The rich arc system produced by the massive cluster of galaxies Abell 1689, as shown by the Advanced Camera for Surveys on board *Hubble*. The shape of its gravitational distortion may be traced simply by eye when the image distortions are so strong and widespread.(NASA, N. Benitez (Johns Hopkins University), T. Broadhurst (Hebrew University), H. Ford (JHU), M. Clampin and G. Hartig (Space Telescope Science Institute). G. Illingworth (Lick Observatory), the ACS Science Team, and ESA.)

Fig. 5.14. The galaxy cluster Abell 2218 and its system of lensed arcs, the first to be recognized as a whole collection rather than as an individual oddity. The splitting of two arcs around the galaxies at lower right indicates their gravitational influence, above and beyond the total cluster's effect. Among the smaller, faint arcs are images of some of the most distant galaxies now known. This *Hubble* image was obtained with the Wide Field Planetary Camera 2 in January 2000, shortly after the third servicing mission. (NASA, Andrew Fruchter and the ERO Team (STScI), Richard Hook (ST-ECF), and Zoltan Levay of STScI.)

detail. A group of astronomers based in California (except for lead author Max Pettini from the UK) used one of the 10-meter Keck telescopes on Mauna Kea for a detailed study of the blue and yellow spectral regions (Fig. 5.16). The scientific interest in this object was so high that an Italian group observed it deeper into the blue and violet range using the European Southern Observatory's Very Large Telescope in Chile; from the latitude of Cerro Paranal at 24.5° south, MS1512+36 never rises as high as 30° above the horizon – not a direction in which one would normally choose to measure blue and near-ultraviolet light. However, at the moment we know of no other distant lensed system for which such detailed studies are possible at all, warranting the extra effort and investment of observing time.

Fig. 5.15. A *Hubble* closeup of the lensed high-redshift galaxy MS1512-cB58, seen as the slightly curved bright object at center. The foreground galaxy cluster gravitationally distorts its image into a section of an arc, making the whole object perhaps thirty times brighter than we would see it without the cluster. This *I*-band image, in the near-infrared, is the sum of sixteen exposures totalling 5.5 hours' integration time. (Data from the NASA/ESA Hubble Space Telescope archive; original principal investigator Howard Yee.)

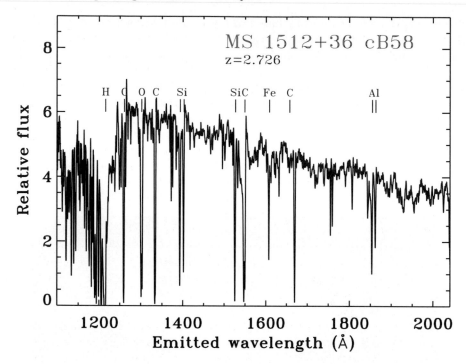

Fig. 5.16. Keck Observatory spectrum of MS1512-cB58 , showing numerous absorption lines from stellar winds and interstellar gas. The gravitational amplification of the brightness of cB58 by the foreground cluster allows us to study it in greater detail than any other known galaxy which is so distant and thus seen at such an early cosmic epoch. The atomic species responsible for some of the strongest spectral features are marked. (Courtesy Charles Steidel, Caltech/ W.M. Keck Observatory.)

An international team led by Richard Ellis and Jean-Paul Kneib combined the imaging quality of *Hubble* and the light-grasp of the Keck telescopes to uncover the kind of object astronomers have long thought might represent the early building blocks of today's galaxies. Comparing the intensity of lensed arcs in various filters to select the ones most likely to be at large redshifts, they found a tiny object at $z = 4.5$ – seen roughly 90% of the way back to the start of cosmic time. This object can be observed only because of the gravitational amplification, which makes it appear thirty times brighter than it would otherwise. Then, closer examination of the region of sky where Abell 2218 would lens most strongly showed a galaxy at $z \approx 7$, studied with a combination of *Hubble* optical images and *Spitzer* infrared detections. This galaxy shows three images, two of which are large, symmetric, and located in the right place to be images of such a distant object. This galaxy is particularly interesting, as many of its stars would have formed so early that they contributed to the heating and ionization of the intergalactic gas which continues to the present day. Efforts continue to extend such searches, especially extending into the infrared where we see the light from the most distant objects.

Astronomers are becoming discerning customers for gravitational telescopes. Comparison of X-ray data on the hot cluster gas with the distributions of galaxies in clusters shows that we see the most lensed arcs, and thus the highest magnification, for clusters which have more than their share of smaller knots of galaxies. Such knots can persist for a long time after a group falls into a cluster, or two clusters merge. Seeing two or more such knots close together essentially adds their lensing effects, making these particular clusters more effective than their mass alone would suggest. Thus, detailed studies of galaxy clusters can not only tell us about the clusters and their history, but show us the most promising places to look deep into the early Universe.

The gravitational telescope can also serve us even when it operates so weakly that nothing looks odd. Any sufficiently distant object will be distorted and amplified (or de-amplified) by the combined, slight effects of whatever masses lie near the light path toward us. In fact, this sets a limit to how precisely we can be sure we see such distant systems. For an individual distant galaxy, it may not matter much, as we will not have a way of knowing that we see it stretched by 3% at a certain angle. However, if we can measure enough galaxies in a certain region, we can tell whether there is an average stretch in a certain direction more strongly than would be expected from chance. This is the realm of *weak lensing*.

Detection and exploitation of weak gravitational lensing has proven to be a challenge both for observational technique and computational efficiency. The effect can be so weak that it is easily masked by slight changes in the telescope's image quality from point to point in the image. The requirements are so stringent that reference stars must be used to measure and remove residual image elongation, to leave the minute signal that all galaxies in a certain part of the sky have a particular direction of elongation. Since galaxies are randomly projected on the sky, and have a range of shapes to begin with, weak-lensing studies must entail huge samples of galaxies to average these distributions. Making this technique work had to await large-format digital detectors, to cover a large region at once and be able to compare everything in that field; and fast enough computers for the considerable level of data analysis needed. One of the first successes in this area was reported in 1990 by Tony Tyson, of what was then Bell Labs, along with R.A. Wenk and software guru Frank Valdes. They looked at two very rich and massive clusters (including Abell 1689, familiar from its very rich system of strongly lensed arcs, as in Fig. 5.13) from Cerro Tololo and Mauna Kea. To reduce artifacts from the detectors, they built up images from large numbers of short exposures with small telescope motions in between, so that any residual systematic problems that were not fully removed in calibration would average out when the images were stacked. They then modelled and removed the bright galaxies in each cluster from the data, revealing numerous background galaxies. The colors of these galaxies give an estimate of their redshifts, important since more distant galaxies will be more strongly lensed. Finally they could automatically identity background galaxies in various redshift ranges and measure the orientation of each image, seeking large-scale patterns. Their detections showed that dark matter was distributed throughout the cluster, over roughly the same region as the visible galaxies, and provided the first direct "images" of the dark matter.

Weak lensing has become a bigger business by now, bolstered by two instrumental developments. Of course, *Hubble's* high resolution will reveal more subtle distortions, and in more distant galaxies, than we can see from the ground. It has been used both to look at the environments of known clusters, mapping the dark matter, and in a parallel mode to randomly sample the weak lensing all around the sky. This kind of observation is almost free (costing essentially the overhead for staff to schedule the observations and support for the extra telemetry and archiving). *Hubble* was built with the capability to use more than one of its science instruments at a time, since it was clear that its images of even random spots in the sky would have considerable scientific value for survey programs. Such parallel exposures have been important in weak lensing, in finding faint multiple-image gravitational lenses, and in tracing galaxy evolution in the distance range where the Deep Fields are too small to include many galaxies.

A recent result from the Sloan Digital Sky Survey uses a novel weak-lensing application, the measurement of which does not even require measuring image distortions. A team of seventeen astronomers, led by Ryan Scranton of the University of Pittsburgh, correlated the properties of a uniformly and quantifiably selected set of 200,000 quasars with locations of filaments of galaxies stretching between clusters. Cosmological simulations suggest that these filaments are dynamic structures, in which mass is flowing toward clusters, and that the galaxies there should be accompanied by enough dark matter to provide a net amplification of distant sources. Indeed, the quasars located along lines of sight through these filaments are slightly brighter than those elsewhere, and the effect changes with the redshift of the quasars just as expected for gravitational lensing. Low-redshift quasars are almost unaffected, since the "lever arm" for redirecting their light is too short, while the effect increases for larger redshifts. This kind of study is possible only because the Sloan survey delivers data samples which are large, reliable, and robustly calibrated. There is already planning for a yet deeper survey instrument, in the 6-meter class, which would repeatedly scan the entire visible sky in search of moving or varying objects. Going beyond these aims, it would yield powerful results on non-moving targets as well. Stacking images from multiple observations of the same patch of sky in such a survey would provide a yet richer digital sky for weak-lensing analysis, giving us the ability to map mass in the Universe solely from its distortion of spacetime.

5.7 Gravitational microlensing – back to stars

Eighty years after Einstein's original (and unpublished) calculations about lensing effects on starlight, it finally made sense to return to the situation he first considered – transient, chance alignments of two stars from our vantage point. It was this simple situation of a point light source and point mass that the original equations described, and that Sjur Refsdal had started with in the 1960s. When he factored in the precision of the alignment needed for us to see a random star gravitationally magnified by another random star passing in front of it, Einstein concluded that the effect would be practically unobservable; we would need to keep tabs on millions of stars to see even one event every few years. But by 1990, not only was the instrumental and analysis machinery up to this task, but there was a pressing reason to actually carry out such a demanding experiment. Its results could bear on one of the most pressing problems in astrophysics: the nature of dark matter.

For years (in retrospect, since a paper by Fritz Zwicky published in 1938), evidence had mounted that all was not well with our understanding of motions within and between galaxies. Zwicky had noted a huge discrepancy between the motions of galaxies in the rich Coma cluster – motions which must be in balance with the cluster's overall gravity if it is not to either collapse or explode, and the mass we could account for in the galaxies. Differences in their Doppler shifts tell us that individual galaxies are moving so fast within the cluster that they would long ago have flown apart, if their motions were influenced only by the amount of mass we could account for in the stars within the galaxies. The shortfall was a full factor of 10 in mass; Zwicky in fact used the phrase "dark matter" to describe the culprit. The problem did not apply just to Coma, or just to clusters of galaxies. It turns up over and over, as data are adequate to reveal it in more clusters of galaxies.

We get the same story from two other completely different kinds of evidence. As we have seen, gravitational lensing gives us a way to measure masses of the lensing objects, galaxies or clusters. The mass required for the distortion we see is always much greater than we can account for in ordinary matter, including allowance for all the white dwarfs, neutron stars, stellar-mass black holes, and cold interstellar matter that we do not see directly but should be there. In addition, X-ray observations reveal vast pools of hot gas trapped in the gravitational wells of clusters. This gas is so hot that it would have long ago streamed out of the clusters unless it is being kept in place by gravity much stronger than the visible material can provide. All three techniques – galaxy velocities, gas temperature[5] and gravitational lensing – measure nearly the same amounts of this unseen material.

Closer to home, individual galaxies show evidence for dark matter, including the Milky Way. Efforts to estimate the mass in galaxies date to the first spectroscopic measurements good enough to show the rotation of spiral galaxies. Spiral galaxies are particularly amenable to this kind of investigation, for reasons both practical and theoretical. Since the spiral arms are marked by regions of star formation, and these regions are full of ionized gas lit up by hot stars, they funnel

---

[5] These first two are independent, but rely on the same physical principle. Using the X-ray gas replaces galaxies with individual subatomic particles moving in the same gravitational environment, with the temperature as a measure of their speeds. The X-ray approach has the advantage that, while you can run out of galaxies to measure for increased statistical certainty, you can always take a longer exposure and get more X-rays to analyze.

much of the stars' ultraviolet light into a few strong peaks in the visible part of
the spectrum. Such peaks can be very strong indeed – much easier to measure
than the weak absorption dips from the atmospheres of the stars themselves,
and functioning as wonderful signposts to use the Doppler shift to measure the
velocity differences from place to place in a galaxy. In addition, the spiral arms
in a galaxy lie in a flat plane and contain stars and gas moving in near-circular
orbits, much like those of the planets in our Solar System. And likewise, we can
use a technique to measure the mass of a galaxy similar to the first way of measuring the Sun's mass – one that traces back to Johannes Kepler's three laws of
planetary motion discovered early in the seventeenth century. Three quarters of
a century later, Isaac Newton developed the mathematical techniques to show
that these three laws followed from his law of gravitation (a law which is a very
accurate approximation of the results of general relativity as long as the velocities are not too high and the masses are not too dense). In particular, if we
look at Kepler's Third Law, which states there is a constant of proportionality between the square of any planet's orbital period and the cube of its mean
distance from the Sun, Newton tells us that the numerical constant has a deep
meaning – it tells us the Sun's mass. Different mass, different constant.[6] Kepler
himself had been able to work from copies of Galileo's sketches and show that
Jupiter's moons satisfy this same law as they orbit the giant planet, with a different numerical constant (one that encoded the mass of Jupiter). Until the era
of spacecraft flyby missions, this was how we could best measure the masses of
distant planets, and the reason it was such a breakthrough to discover Pluto's
moon and resolve important questions Pluto's origin and nature.

So, to measure the mass of a spiral galaxy, we set the spectrograph slit along
its major axis, and measure the differences in Doppler shift from place to place
moving outward from the center (Fig. 5.17). We need to pick a galaxy that is
not too nearly face-on, so the Doppler shifts will be significant, but not too
edge-on, because then we have trouble seeing through its own dust and only
see the outer regions. After correcting for this viewing angle, which reduces the
velocity differences we see, we obtain a map of orbital speed with distance from
the center. A similar map for the Solar System is quite simple, and was known
to Copernicus – a smooth drop in orbital speed with distance. This form, later
to be called Keplerian, tells us that the mass in the Solar System is strongly
concentrated to the center. For mass that is widely distributed, like a galaxy,
we can start by using one of Newton's theorems, dramatically simplifying the
calculations. Objects orbiting inside a uniform, spherical shell of matter do not
respond to its gravity, because the net attraction of all its pieces exactly cancels.
So, to the extent that a galaxy's mass is spherically distributed, the speed of
an orbit tells us how much of the mass lies inside that orbit, and increasingly
larger orbits tell us how much mass lies within each of them, and so not only
how much mass there is, but where it is found. As we move outward, we should

---

[6] To be slightly more exact, the constant tells us the sum of the masses of the Sun
and the planet in question. The difference amounts to 0.1% for Jupiter and much
less for any of the other planets in our system.

Fig. 5.17. Rotation curve of the nearly edge-on spiral galaxy NGC 5746, aligned with an image of the galaxy. The points show the relative Doppler shift along the spectrograph slit; the left side of the image is the receding side of the galaxy disk. After an initial rise in orbital speed outward from the center, the orbital velocity remains roughly constant – a flat rotation curve. Spectroscopic data from the 2.1-meter telescope of Kitt Peak National Observatory; image from the 1.1-meter Hall telescope at Lowell Observatory.

eventually see the orbital speed drop with radius (as in the Solar System), for orbits outside most of the galaxy's mass.

Over and over, the same answer emerged, although it took some time to sink in. With hindsight, the velocity pattern in our own Galaxy should have given us a clue. However, our initial expectations fit early rotation-curve data well enough to have left a lasting (and wrong) impression that we already knew something about galaxy masses. This changed first with new technology providing superior data, and with a mental framework to describe it. The crucial data, for most astronomers, built up over several years in the form of rotation curves measured with a new image intensifier. Vera Rubin and her colleagues at the Carnegie Institution of Washington used such a device to measure Doppler shifts farther into the faint outskirts of spiral galaxies than previously possible, expecting to determine their total masses and constrain the population of faint stars which contribute almost nothing to the overall light. But as they measured the spectra, the orbital speed stubbornly refused to fall as they look farther and farther from the galaxies' cores. The speed remained nearly constant with radius, described as a *flat rotation curve*, illustrated in Fig. 5.17. Not only were the orbital speeds too fast to be held in place by the gravity we knew how to account for, but the pattern showed that each galaxy was surrounded by an invisible halo, comprising most of the galaxy's mass, and much "fluffier" than the galaxy's starlight. A galaxy is a candle floating in a vast dark pool.

In the mid-1970s, Rubin's group published dozens of such rotation curves, pointing to a major gap in our understanding of the Universe. (For some years, the preferred term was "missing mass", rather than dark matter. However, it was not the mass that was missing, but our understanding.) A widely-cited 1979 review article by astronomical pundits Sandra Faber and Jay Gallagher presented this evidence systematically, and played a large role in the community's recognition of the existence and importance of dark matter.

The speculation which set in as to the nature and origin of the dark matter has scarcely subsided, although we do know enough to help channel the debate by now. There have also been explanations proffered which do not involve unseen mass. Mordehai Milgrom, the theorist who was one of the first to work out the jet dynamics in SS 433 (Chapter 2), has championed the notion that the fault lies not between the stars, but between our ears. He points out that our first-hand knowledge, where gravity behaves as Newton (and, even better, Einstein) describes, is limited to size scales much smaller than galaxies. We can track objects very well indeed within the Solar System, and up to globular clusters tens of light-years across, and all is well. But if we consider a system much larger, where the acceleration due to gravity (or spacetime curvature) is vastly weaker, we do not yet have independent confirmation that these gravitational theories are accurate. Milgrom proposed a Modified Newtonian Dynamics or MOND, which was designed to make the minimal changes to classical theory needed to produce flat rotation curves strictly as a result of some new physical limit. He started with Newtonian gravity for mathematical convenience, later exploring ways that such a formulation could satisfy the same principles as general relativity. His ideas are still under exploration, particularly whether there is a generalization which gives the gravitational lensing we observe. Looking back, perhaps the surprising thing is that it took so many years for someone to seriously consider whether the problem was in our understanding of the content of the Universe, or its governing law.

In the early 1980s, there was much discussion of whether the dark matter could be composed of WIMPs – Weakly Interacting Massive Particles. Such particles, perhaps even a flavor of neutrinos with significant amounts of mass, were expected in some theories trying to explain patterns seen among elementary particles. We could look for such particles anywhere, if they are widespread throughout our galaxy. We would be as likely to coax one into interacting with our equipment in a laboratory as by sending the equipment into deep space. Most cosmologists would still give their best odds on the idea that dark matter is some kind of particle produced in extravagant abundance in the early Universe. But it was not at all clear, especially at the start of the 1990s, that a different kind of object altogether might make up the dark matter. If not WIMPs, perhaps the Universe was populated by MACHOs. These Massive Compact Halo Objects could include stellar-mass black holes, white dwarfs or neutron stars, rogue planets wandering free of stars, or any other exotic collection of matter, as long as it was gravitationally held together into a single object. The interesting thing about MACHOs is that we could expect to see them, if such things accounted for much of the dark matter. Such objects could not help being detected by gravitational lensing, if we could watch enough background stars long enough.

We even had a good idea of what to expect, and how to use the answers to refine our understanding of the stars that are too faint to make much difference in a galaxy's light. For a certain number of stars at a particular distance, we can calculate how often one lensing mass moving at a particular distance and velocity should lens a background star. Then we know how often to expect lensing for any particular density of lens masses. The signature we seek here is not image splitting, but the transient gravitational amplification in brightness (the same at all wavelengths) of a background star as an unseen lens mass drifts in front of it. This amplification might last from hours to months, depending on the particular set of distances and velocities involved.

This kind of lensing, in which a single star temporarily amplifies the brightness of a background object as it moves across our line of sight, has become known as gravitational microlensing (more or less because the splitting into multiple images would give separations of microarcseconds, as well as being on a much smaller scale than the familiar lensing by whole galaxies). After some theoretical developments in the 1960s, particularly by Sjur Refsdal, it entered the astronomical toolkit by a back door – through our wondering whether it could explain some oddities reported in the statistics of quasars.

As with other astronomical objects, much of our understanding of quasars hinges on our knowing their distances, in this case usually derived from the Hubble law relating redshift to distance (Chapter 8). This is the kind of calculation which told us that many quasars not only outshine any normal galaxy, but the light of hundreds at once, despite the tiny sizes indicated by reverberation mapping (Chapter 2). The problems in understanding the energy densities of quasars led some theoreticians in the 1960s to wonder whether they might somehow be closer and smaller than we thought. Even though this particular problem has gone away with the realization that there are Seyfert nuclei in very nearby galaxies with just the same problem, a few astronomers have argued on more empirical grounds that all is not as it seems with some quasars.

Halton (Chip) Arp, a long-time staff member at Mt. Wilson and Palomar now working in Munich, has pointed out numerous examples of quasars located suspiciously close to the directions of bright galaxies, which would have to be much closer to us in the conventional interpretation. Beginning in 1966, he presented examples of such associations, sometimes involving two or three quasars, often with a disturbed or active galaxy (Fig. 5.18). If these are indeed genuine, physical associations, some large part of the redshift of these quasars arises not from cosmic expansion, but from something new to our understanding of physics. The controversy has hinged on whether these associations are statistically significant, since quasars surveys were, at that time, very inhomogeneous over the sky, making it unclear whether any suitable control sample existed. The expected number of chance associations is very sensitive to the mean number of similar quasars all

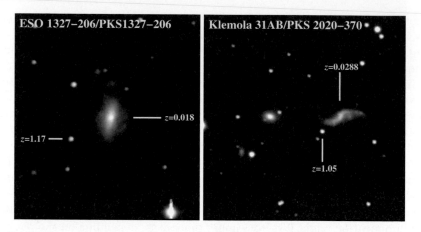

Fig. 5.18. Two of the galaxy–quasar pairs presented by Arp as evidence that quasar redshifts might not be what they seem. The galaxy ESO 1327-206 shows the kind of tails and loops of stars that would result from a merger; one of these loops crosses the quasar's line of sight. Klemola 31 is a small group of galaxies; the quasar PKS 2020-370 appears closest to the elongated spiral. Images obtained with the 3.6-meter telescope of the European Southern Observatory, La Silla, Chile.

over the sky – a number which has usually proven to be underestimated because of selection biases which plague many ways to find quasars.

Still, what was not in dispute was that Arp had presented examples of quasars at high redshift, appearing in our sky next to galaxies of modest redshift which could appear quite normal. As soon as lensing of quasars by galaxies had been demonstrated, the idea arose that perhaps some of these quasars were gravitationally amplified, not by the galaxies themselves, but by stars or clusters within the galaxies. This solved Arp's statistical challenge because of a property that quasars share with stars and galaxies. There are many more faint quasars than bright ones. In fact, for a wide brightness range, the number grows as we look fainter even faster than one expects simply from probing a greater volume – one of the prime pieces of evidence that quasars have evolved. This means that if one of one hundred quasars is amplified in brightness by, say, a factor of ten, as its light passes near some star in a distant galaxy, it will have a huge effect on the statistics of its new bright peers, because they are perhaps a hundred times rarer on the sky. Claude Canizares at MIT was examining this possibility by 1980, particularly whether globular star clusters around these galaxies might be responsible.

Having been taking photographs with Lick Observatory's venerable Crossley telescope and prowling through its archive of plates taken by earlier observers, I saw an opportunity to test this idea further. All other things being equal, a more massive lensing object will produce a longer-duration lensing event. Since they appear in the same fields as relatively bright galaxies, 22 of these quasar–galaxy pairs had been photographed with the Crossley as much as 80 years earlier. I reobserved them (Fig. 5.19), using a plate-filter combination intended to mimic the sensitivity of the old blue-sensitive emulsion and mirror coated with silver (instead of the contemporary aluminum coating). In every case, the quasar had been bright already early in the century, and none were more different in brightness than one might expect from the variability of typical quasars. If these objects were microlensed, they had remained bright for so long that the bodies responsible were not ordinary stars – they were distributed throughout an extensive halo and, after a flurry of calculations done in the unusual comfort of house-sitting a large Victorian dwelling, I found that they must have had at least 10 solar masses. This in itself might fit for the collapsed black-hole remnants of the first stars, but, as Jeno Barnothy was quick to point out, the number of such objects needed to see so many instances of lensing was implausible even if they comprised dark-matter halos by themselves. There was always the possibility that these quasars were not microlensed... As a naive graduate student, I sent this report to *Nature*, where Canizares' original suggestion had appeared, and which summarily rejected it. Taking sage advice from my thesis adviser Joe Miller, I put the report away for a couple of weeks, took a deep breath, changed the British to American spelling, and sent it to the *Astrophysical Journal Letters*. It appeared as the lead paper for August 1, 1982, after receiving a glowing referee's report. I continue to tell graduate students about this episode, as evidence that the ways of peer review can be very mysterious.[7]

Gravitational microlensing has occasionally resurfaced in the galaxy–quasar context, but its most effective use has been between stars (as Einstein considered). The possibility that MACHOs could be detected, and at the very least we would know how much mass was tied up in very dim stars and stellar remnants, led to several large-scale projects starting in the early 1990s, designed to monitor the required millions of stars to see at least one such event per year. The most ambitious was the MACHO project, led by New Zealand native Charles Alcock of the Lawrence Livermore National Laboratory in California. That affiliation was a sign of a new marriage of techniques; particle physicists were already accustomed to work in large teams on projects requiring management of enormous data sets across geographically dispersed groups, so it was natural for LLNL to be involved here, and Fermilab to be an integral part of the Sloan sky survey.[8] The MACHO team refurbished a 1.3-meter instrument, the Great Melbourne Telescope, at the Mt. Stromlo Observatory outside Canberra.[9] The southern location gave the project access to the best celestial real estate to seek microlensing – the central bulge stars of the Milky Way and the stars of the Magellanic Clouds. The telescope was upgraded and equipped with a large-

---

[7] And I do not except peer review when I practice it.

[8] And after managing this group, Alcock was chosen in 2004 to direct the Harvard–Smithsonian Center for Astrophysics, one of the largest and most diverse astronomical institutions on the planet.

[9] This telescope was destroyed, along with almost all the other Mt. Stromlo facilities, when an intense brush fire swept the site in January 2003.

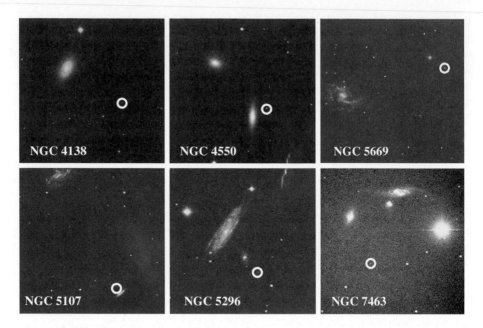

Fig. 5.19. Photographs of some of Arp's quasar–galaxy pairs, obtained with the 0.9-meter Crossley reflector circa 1981. These photographs repeated serendipitous observations with the same telescope as much as 80 years earlier, showing that these objects were not short-lived lensing events. Names are for the brightest galaxy in each frame; quasars are circled, and in some cases quite faint.

format CCD camera, and connected for rapid data transfer. After a start to data taking in 1994, the collaborators were ready in 1998 to announce the detection of microlensing events. First they had to identify variable objects, and remove the great majority which were ordinary kinds of variable star. A pointlike mass acting on the light of a small background object produces an amplification history whose form depends only on the miss distance of the light (compared to the size of the nominal Einstein ring) and the relative transverse speeds of source and lens (again expressed in terms of the Einstein radius, this time in how long it takes the stars' relative separation from our vantage point to crossing it). Furthermore, this behavior is exactly the same at all wavelengths, which is why the MACHO project took data in both red and blue filters. After a decade of observations, the MACHO team has detected more than 400 probably lensing events. Some of the best-measured events are shown in Fig. 5.20, as time-series plots of each star's brightness around the time of peak amplification.

At the same time, other groups had been using variations on this approach to seek microlensing, with different choices of field of view, detection strategy, and analysis. A certain theme seemed to pervade their naming, with MACHO being joined by OGLE (Optical Gravitational Lensing Experiment, led by a Polish group, observing from Chile) and EROS (Expérience pour la Recherche d'Objets Sombres, with a French team also using a telescope in Chile). These

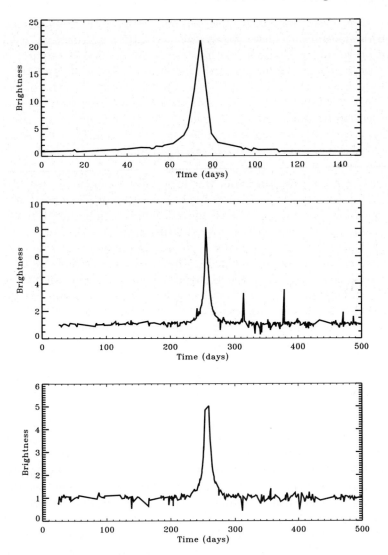

Fig. 5.20. Three of the highest-amplification gravitational microlensing events found by the MACHO project. These are all stars in the Large Magellanic Cloud. The red-light measurements are shown, scaled to each star's normal brightness. The measurements are connected by lines for legibility, although missing or noisy data introduce asymmetries in some places which are not implied by the data. These events have peak magnifications from 7–47, and timescales near 400 days. (These public-domain data were obtained by the MACHO Project, jointly funded by the US Department of Energy through the University of California, Lawrence Livermore National Laboratory under contract No. W-7405-Eng-48, by the National Science Foundation through the Center for Particle Astrophysics of the University of California under cooperative agreement AST-8809616, and by the Mount Stromlo and Siding Spring Observatory, part of the Australian National University.)

teams erected dedicated 1.3–1.5 meter telescopes to continue these studies more effectively in the long term. These surveys all agree that microlensing events are seen; that there are perhaps more than the simplest models require, but not so many as to require significant new populations of objects; and that it is difficult to determine the exact locations and masses of the lensing objects. The latter problem is especially galling, and comes about because of the very simplicity of lensing by point masses. The lensing brightness behavior tells us only a combination of the size of the Einstein ring, which combines its distance and mass, and the transverse velocity of the lens relative to the background star. If we do not know where the lens is along the line of sight, its mass is indeterminate. Sometimes we can break this impasse – for example, when the lens is a star later seen directly as it moves out of the way, or when the Earth's orbital motion carries us through the lensing event at a changing rate. But there is still argument about *where* the lenses are, and therefore *what* they are. Lenses in front of the Galactic Center might be stars simply on the front side of the central bulge we are observing, or might be quite nearby ones belonging to our galaxy's disk, or some kind of genuine halo object in between. Lenses seen toward the Magellanic Clouds might belong to either galaxy. This uncertainty plagues analyses of the overall microlensing results.

To be sure, microlensing can do more than tell us the masses of unseen intervening objects. As more data served as fodder for examination, theorists considered the signatures of lensing by binary stars, the effect of details on a star's surface, and whether we might detect even Earth-mass planets by their perturbations of the lensing caused by the parent star. One faint red star was found by *Hubble* several years after it lensed a distant star near the Galactic Center, once it had moved free of the glare of the distant star. That gave this unassuming star the distinction of being only the second star (after the Sun) outside of a binary system for which we had measured a mass directly. Analysis of such subtle microlensing effects is now a rich field of inquiry. In 2003, the OGLE team reported the first microlensing detection of a planet, from two spikes in the amplification history of a star which match the expected effect of a planet crossing the Einstein radius. This was confirmed by data from the MOA project (in New Zealand, of course,[10] although the acronym stands for Microlensing Observations in Astrophysics). And in May 2005, a global team led by Andrew Gould of Ohio State University reported clear detection of a lensing event involving both a star and a planet, with only a few times the Earth's mass. If such events can be detected in real time, further information is available from a wide variety of observatories. Two contributors to this event were amateur astronomers in New Zealand, working with 25- and 40-centimeter telescopes.

Looking farther afield may be able to resolve the ambiguities left from the MACHO project and similar experiments. If we look at microlensing within the Andromeda galaxy, or even between stars in two different distant galaxies. the combination of mass and velocity that we measure can become different

---

[10] Home to the wide variety of giant bird species known as moas, none of which have been reliably seen since the arrival of Europeans in the eighteenth century

enough that an analysis of all the surveys should give us typical lens masses for the whole set of galaxies. Twenty years after investigating quasar–galaxy microlensing, I rejoined the gravitational-lensing game in such a project. This came about through quite a different research interest: the role of dust in galaxies. Astronomers knew that spiral galaxies contain dust before they were quite sure that they were galaxies. The importance of the dust in what we see – how much starlight it blocks, whether it changes the galaxy's overall appearance, how much we need to worry about its reddening effects when comparing galaxies – has been contentious. Some studies have reached the comforting conclusion that it is picturesque but does not change things much, while others have found that we see only a fraction of the starlight from a typical spiral (the lost energy heating the dust grains and emerging deep in the infrared). Some of these studies rely on modelling which has to fold in poorly-known details of galaxy structure (such as the characteristic thickness of a galaxy's disk as expressed in stars of various temperatures). Colleague Ray White III and I embarked on a more direct approach, goading each other into it by our mutual stupefaction at one particular study whose conclusion, if taken literally, would have prevented our being able to see enough galaxies to reach those conclusions.

Knowing of anecdotal cases of one galaxy backlit by part of another, we carried out a more systematic search for overlapping galaxies. In these cases, if the background galaxy is smooth and symmetric, a major unknown drops out of the analysis; all the backlight comes from behind the dust. This project led us to observatories in Arizona and Chile, and eventually to get a program scheduled on *Hubble* to examine the fine structure of dust in the handful of galaxy pairs which our ground-based data showed were most symmetric and most promising for further analysis. Among these is the spectacular system known as NGC 3314. Seen in a cluster of galaxies in the constellation Hydra, NGC 3314 consists of two spiral galaxies almost exactly superimposed; their centers are less than 2 arcseconds apart in our view. While this complicates our analysis for some results, it offers the unique chance to view backlit dust in spiral arms deeper into a galaxy than any other case we knew. The *Hubble* images did not disappoint us, allowing us to derive a radial map of dust absorption throughout the galaxy as well as study its fine structure down to tens of light-years. The spectacle of this image caught other eyes as well. Lisa Frattare, a former summer research student with me, had moved to the Space Telescope Science Institute and joined the Hubble Heritage team at its inception. The team produces a new image every month, from new and archival *Hubble* data, selected for aesthetic impact rather than purely scientific value. When Lisa brought it to the team's attention, the Heritage project decided to make an image release of NGC 3314, which would include obtaining additional data for more realistic color rendition. We had the chance to recommend details of the new images, for best compatibility with our earlier data, especially maintaining the same alignment as nearly as the guidance system would allow over the year-long time between images. White and I gleefully added the new data to our previous harvest in publishing the analysis, while the Heritage team examined the best way to display the color-composite product of the whole data set.

We traded image versions, considering how to display the images for maximum detail and accurate color rendition. In late March 2000, weeks before the release date for the final image, Lisa emailed about something appearing in the new images which had not been there the year before, when our original images had been obtained. It stood out in the color-composite image as green – not because it had that rarest of astronomical colors, but because in the RGB addition of filtered images to make a color version, it had been there for the green image, for half the exposures of the blue image, and not at all for the red data. A bright starlike object had appeared in NGC 3314 some time in the preceding year. This was too bright for an ordinary nova outburst, but faint enough that it could be a supernova only on the rise or (more likely) in its gradual fading afterward. A call for any additional observations failed to elicit any additional detections, so we still cannot be sure what this was. A more remote possibility, which I had briefly considered but dropped, was brought back to mind from some queries sent to the Hubble Heritage web site by alert web readers: why do we not see gravitational lensing between the two members of this galaxy pair? (We probably do, but having no standard of comparison, do not notice the percent or two stretch of the background image.) This brought back to mind the issue of microlensing between stars in the two galaxies. This gave a long distance for rays to be deflected, and a very large Einstein ring size, because even the foreground galaxy is 140 million light-years away.

However, a back-of-the-envelope calculation suggested that we would have to be incredibly lucky to see such a strong amplification from microlensing. There should be a microlensing event going on strongly enough to double the brightness of some star or other in NGC 3314 all the time, but *Hubble* would be unable to detect typical events even for red giant stars. And worse, an event strong enough to produce the mystery object would last only a couple of hours (and we knew this one had lasted at least an hour and a half). So I wrote off microlensing as the explanation for this object, and said so in a piece on the Hubble Heritage web site.

Perhaps I had been too hasty. A year or so later, Dave Bennett, who had been part of the original MACHO collaboration, convinced me that my original back-of-the-envelope estimates of expected microlensing events in the NGC 3314 pair was too pessimistic, by at least a factor of 10. Taking into account what we knew about the mass distributions in the galaxies from mapping their rotation patterns, and our best estimates of the two galaxies' distances, it was in fact reasonable to detect a few such events per year. Not only reasonable, but potentially quite informative. A perennial issue in interpreting these microlensing studies has been the fact that, from the magnification and duration of a lensing event, we measure only a combination of the mass and transverse velocity of the lensing mass. This has led to controversy as to what and where the lenses are. For a distant galaxy pair, the statistics should favor seeing quite different relative velocities, so that we could break this connection to the extent that we are seeing similar kinds of stellar masses in our own galaxy and the members of NGC 3314. Bennett pushed for a very ambitious *Hubble* project, which we proposed and which was approved in 2002. This project involved thirty observations of NGC

3314, in two filters each to check for the color independence which characterizes gravitational lensing. They were spread over a two-year span, with uneven time sampling so we could detect events as short as a week and as long as a year. Each of these observations shows fainter objects, and at higher resolution, than our entire earlier data set (Fig. 5.21).

Fig. 5.21. The spectacular superimposed galaxy pair NGC 3314, as seen from a single *Hubble* observation using the Advanced Camera for Surveys (ACS). Thirty pairs of similar images have been acquired to measure gravitational microlensing of stars in the background galaxy by stars, stellar remnants, and other compact masses in the foreground galaxy. (NASA/ESA Hubble Space Telescope data, from a collaborative project led by D. Bennett.)

The data analysis is still ongoing, refining procedures to match the images at the earliest possible processing stage and reduce the many "false alarms" from instrumental effects, but some objects do appear, brighten, fade, and disappear in a way consistent with being microlensing events (Fig. 5.22).

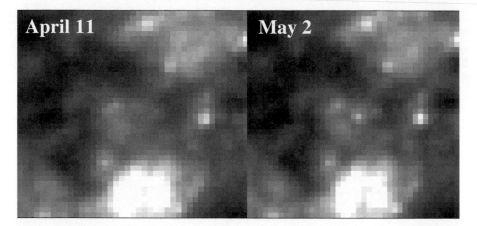

Fig. 5.22. A candidate microlensing event in NGC 3314, seen 6.6 arcseconds from the center of the foreground galaxy. These sections of ACS visible-light images are 2 arcseconds on a side, comparing the same region before and at the height of the brightening event. The peak magnitude of this event, as sampled by the *Hubble* images, reached 27 – a level virtually impossible to measure with any other instrument. Dates of observation are in 2003. (NASA/ESA Hubble Space Telescope data, from a collaborative project led by D. Bennett.)

Meanwhile, groups at Columbia University and in France (under the acronym AGAPE, which is presumably less urgent than EROS) has been looking for microlensing in the Andromeda galaxy, M31. The aim here is to see whether there is a difference between the numbers of events on the near and far sides of the galaxy disk, which would suggest that the lenses lie in a large halo rather than in the disk itself. Arlin Crotts at Columbia University has led a long-term study of Andromeda, to seek evidence of these lensing events. These studies in distant galaxies suffer from a problem which is much less important for nearby lenses – we cannot see the lensed star by itself. We can find the events when lensing brightens a star, but when it fades into the background of the other 10,000 or so stars sharing its space on the detector, we do not know how bright it normally is. Thus, in this case known as pixel lensing, we have one more free parameter and correspondingly larger uncertainties on the values we get back.

As these studies continue, we see gravitational lensing and microlensing becoming mature astrophysical tools, and a powerful implicit window that breaks the privileged position of objects that shine. We can now sample the constituents of the Universe, dark or bright. Eventually, this sampling may extend to still other constituents of the Universe. A few forward-thinking scientists have explored whether gravitational lensing could be a tool in the search for extraterrestrial intelligence (SETI). SETI pioneer Frank Drake pointed out that the most ambitious equipment ever proposed is weak compared to the possible gain by using the Sun's gravitational focussing. Placing even a modest antenna along the distant line of the solar gravitational focus could make it enormously sensitive, albeit expensive to repoint. We might also consider whether civilizations elsewhere might use microlensing between stars to extend the reach of beacons designed for less sophisticated civilizations such as ours.

Further reading:

On the search for early building blocks of galaxies, I can shamelessly recommend William C. Keel, *The Road to Galaxy Formation* (Springer-Praxis 2002)
A Java applet to investigate lensing effects on a background image of the user's choice has been provided by Peter J. Kernan of Case Western Reserve University, at http://theory2.phys.cwru.edu/~pete/GravitationalLens/GravitationalLens.html
Figs. 5.8 and 5.9 were produced with this resource.

6 The stars themselves

The stars embody $E = mc^2$ before our eyes. There, I have put in an equation – something that publishers fear will cost a book a large fraction of its potential readers. Keep reading, please!

Astronomers came slowly to the appreciation that there is something to understand about why stars shine, something that is intimately connected to the life cycles of stars and the dizzying variety of stars that we see around us. Throughout the nineteenth century, astronomers were concerned with the details of celestial mechanics, parallax measurements, and the distribution of stars through space. But asking why these stars are there, and why they are so configured, fell into philosophical speculations that respectable astronomers would approach only with a ginger touch. Sir John Herschel's textbook *Outlines of Astronomy*, reprinted as late as 1902, devoted chapters to the motions of stars, the clustering of stars, and even whether the zodiacal light indicates that the Sun is surrounded by a reflection nebula. He devotes a couple of pages to the idea that the Sun's heat is maintained by infalling meteorites, and precisely once even mentions "processes antecedent to the existence of separate self-luminous solid bodies". In its 1934 edition, E.A. Fath's *Elements of Astronomy* shows the first dawning of an appreciation that stars not only shine but change. Fath writes of Cepheid variable stars as being in a "certain stage of their development", and compares a theory of stellar cooling due to Lockyer with Eddington's notion of progressive evolution to lower and lower luminosities, with the distribution of stars in color and luminosity marking possible paths for this development. He also mentions Henry Norris Russell's ideas that somehow $E = mc^2$ and the conversion of mass to energy is taking place, and his estimates of ten trillion (10^{13}) years for the ages of stars on this basis. Not for the last time, astronomers seemed to have a huge discrepancy between estimates for the ages of stars and of the whole Universe. Fath concludes, "For the present, there seems to be no way to reconcile these antagonistic views of the time-scale..."

Uncovering the processes that make stars shine, and evolve, was one of the major steps in astronomy's century-long transition from being a purely descriptive science to an historical one. In the nineteenth century, the only known energy sources which might act in stars were chemical combustion and release of gravitational energy, either through infalling material or a slow contraction of the Sun. These both ran afoul of the timespans being revealed by geology and paleontology (Fig. 6.1). These suggested that the Earth had been here, and plants had been illuminated by sunlight not much different from what we see today, for hundreds of millions of year. Chemical combustion among the most favorable substances releases far too little energy for even a mass the size of the Sun to burn for so long and so brightly; we would then deal with the Sun being younger than recorded human history. Gravitational energy, based on a slow shrinking of the Sun and conversion of gravitational energy into heat, could do better, into the millions of years, but would have put the Earth inside the Sun while dinosaurs walked.

In 1862, William Thompson (Lord Kelvin, after whom the temperature scale is named) presented an influential estimate of the Earth's age, based on estimates of the heat flowing out of the Earth and extrapolating to determine how long ago it would have been completely molten. This extrapolation indicated an age around 100 million years, assuming that no additional energy sources were present. This already presented a problem for understanding geological formations. We later learned, of course, that there is an additional energy source in the Earth's interior which adds heat over time – decay of radioactive atoms. Discoveries in radioactivity led to the growing suspicion that this phenomenon might be related to the Earth's heat, and dramatically revise the results of Kelvin's heat-flow argument. Years later, physicist Ernest Rutherford recalled an encounter with Lord Kelvin, during Rutherford's 1904 lecture on radioactivity given at the Royal Institution. Kelvin was in the audience. In his much-retold account, Rutherford recalled that "to my relief, Kelvin fell fast asleep, but as I came to the important point, I saw the old bird sit up, open an eye, and cock a baleful glance at me! Then a sudden inspiration came, and I said Lord Kelvin had limited the age of the Earth, provided no new source was discovered. That prophetic utterance refers to what we are now considering tonight, radium! Behold! the old boy beamed upon me."

In itself, the kind of radioactive phenomena Rutherford spoke of did not solve the second age problem, one that Kelvin had also demonstrated. If the Sun shines purely by radiating heat stored at its formation, it must shrink and cool over time. By the most generous assumptions, that would have made the Sun larger than the Earth's orbit less than 500 million years ago (and the problem worsens if we recall the existence of Mercury and Venus).

Sometimes we must wait on our understanding of physics to catch up to what we see in the cosmos. In this case, we had to know about the transformation of mass into energy during nuclear fusion. Because fission occurs among natural substances in Earth, it was from this direction that our understanding of these process unfolded.

A year before the first data hinting at the enormous power to be released by fusion, Henry Norris Russell presented a carefully argued discussion [1] of how the mysterious stellar energy source must behave. Russell showed, without recourse to equations, that whatever makes the stars shine, it must become rapidly more

---

[1] "On the Sources of Stellar Energy", *Publications of the Astronomical Society of the Pacific*, August 1919. This article was cited with admiration in John Bahcall's review on the occasion of the award of the 2002 Nobel Prize in physics to Ray Davis.

Fig. 6.1. The evidence that the Sun has shone for a long time, and more or less as brightly as it does today, is encapsulated in these ferns. One set was laid down in the coal-age swamps of what is now Alabama about 310 million years ago. The other set still grows in my back yard, within a half-hour's drive of where the old ones were found. Similar leaves, of about the same size, supported both sets of plants by photosynthesis. Both sets must therefore have grown beneath very similar Suns.

effective at higher temperatures, and that this feature made it possible for stars to be stable over long periods.

Perhaps the first direct hint of just how stars shine came from a series of laboratory measurements by F.W. Aston, published in 1920. Among a large set of measurements of the masses of atoms, he briefly notes the curious fact that four hydrogen atoms outweigh a helium atom by 0.7%, although simply adding the masses of their constituents would make helium more massive by 0.2%. Aston did not explore the question further, but it was taken up soon enough by physicists and astrophysicists alike. Laboratory experiments repeatedly confirmed that relativity's generalization of conservation of mass and energy into the combined conservations of mass–energy was as exact as we could measure – so exact that a breakdown in the bookkeeping reliably signaled the existence of unseen particles. This pointed directly to the potential release of vast amounts of energy. How vast? $E = mc^2$, after all. This drove Eddington to write, in the same year, that "If, indeed, the sub-atomic energy in the stars is being freely used to maintain their great furnaces, it seems to bring a little nearer to fulfillment our dream of controlling this latent power for the well-being of the human race – or for its suicide."

Our understanding of fusion was initially driven by theory, since we could not reproduce or measure this phenomenon as we could the fission of heavy nuclei. Key steps were taken with the quantum understanding of what happened when particles approached very closely, in which a process called tunneling could allow them to interact even when electrical charges of the same sign would (in ordinary classical physics) have kept them too far apart. By 1938, the stage was set for very rapid progress. In Germany, Carl von Weizsäcker set out a catalytic process whereby small amounts of carbon, nitrogen, and oxygen could drive the net fusion of hydrogen into helium. This process, the CNO cycle, is not important in the Sun, but is the dominant fusion path in hotter stars. And in the USA, Hans Bethe identified the so-called $p - p$ chain of reactions that power the Sun and similar stars. In a *tour de force* of theoretical nuclear physics, Bethe set out within a year all the basics that we still recognize for these reactions.

The precise trade between mass and energy in nuclear reactions can be described by the binding energy of an atomic nucleus. This is the amount of energy required to dismantle the nucleus, pulling it apart particle by particle. By symmetry, the same amount of energy must be released when such a nucleus is created from particles that are initially independent of one another. Since we consider what happens particle by particle, we usually deal with the binding energy per nuclear particle. On doing so, we find interesting properties of the so-called curve of binding energy, which compares binding energy per particle for elements of various mass. Such a graph is shown in Fig. 6.2. Any reaction moving upward along either branch of this curve gives off energy; any reaction moving downward requires more energy to initiate than it generates. We see from this that transforming hydrogen to helium is not only the single most potent nuclear reaction for generating energy; by itself it releases 3/4 of all the nuclear energy which can be extracted from a given amount of material. This is why stars spend most of their energy-producing lives powered by hydrogen.

In beginning astronomy books and freshman lectures, it is customary to compare the Sun's energy to an H-bomb (more correctly, a fusion explosive). The basic principle is certainly the same – release of energy by buildup of heavier atomic nuclei from lighter ones, with most of the energy being liberated as protons and neutrons join. However, most of the specific nuclear reactions are different. A thermonuclear bomb (Fig. 6.3) is designed for greatest impact, with the precise conditions and chemistry carefully optimized for maximum energy release from a small and light package. In contrast, the energy release per kilogram in the core of the Sun is tiny – much less than the warmth generated per kilogram of the human body. The Sun is so much larger, and has done this for so long, as to keep it much hotter and brighter (Fig. 6.4).

The scale of a star's interior explains why the first reaction in the so-called $p - p$ chain is one not used by terrestrial bomb designers. This fundamental reaction is denoted by

$$p + p \rightarrow D + \nu + \text{energy}$$

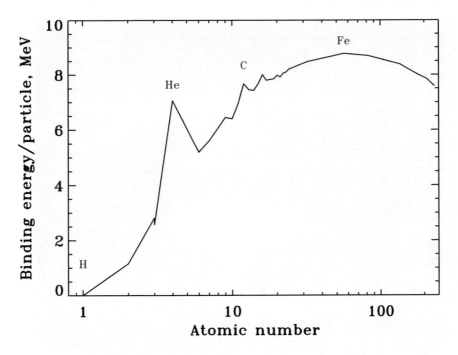

Fig. 6.2. The curve of binding energy for atomic nuclei. As is customary, this is shown in binding energy per nuclear particle, removing the effects of scale in more massive nuclei. The energy is in MeV (millions of electron volts), a convenient unit on nuclear scales. Nuclear reactions which move upward on this plot generate energy, while those moving downward remove it from the environment. Hence, fusion from hydrogen to helium generates energy, as does fission of heavy elements such as uranium into lower-mass products. Iron stands at the peak of this curve, the most tightly bound nucleus. No nuclear reactions involving iron generate energy, a fact which ends the lives of massive stars.

where each p represents a proton, the nucleus of the most common form of hydrogen, and D represents the nucleus of deuterium or heavy hydrogen, in which the proton is joined by a neutron (ν denotes a neutrino, about which more shortly). This reaction requires double nuclear alchemy. First, while the two protons are at their closest and about to be repelled by the electrical force of their like charges, the weak nuclear force can act (once in a great many encounters) to transform one of the protons into a neutron, carrying away the electrical charge in the form of a positron (the antiparticle of an electron). Only then will the strong nuclear force, affecting protons and neutrons, act to glue them into a nucleus. The weak force is quite weak; at the high temperature and pressure in the Sun's core, the typical time for a given proton to be involved in this reaction is close to 5 billion years. Bomb designers start with elements that have already overcome this slow hurdle, including nuclei that can be broken

Fig. 6.3. Humanity does fusion. The Castle Romeo atmospheric nuclear test, an 11-megaton explosion carried out near Bikini Atoll on March 26, 1954. This yield marks the conversion of about 500 grams of bomb material into energy, about what the Sun does in 10^{-10} second. (US Department of Energy photograph.)

into heavy hydrogen isotopes when a fission explosion occurs. This means that the heavier isotopes of hydrogen – deuterium (D) and tritium (T) – are used, passing up the slow steps of their production from protons. Tritium, the heaviest form of hydrogen, has a radioactive nucleus with two neutrons and a half-life of only 12.3 years. Thus, every year sees a 5% decrease in the amount of tritium in any sample, which is a factor in the debate over testing of nuclear weapons – the decay of tritium gives them a specific shelf life and warranty expiration date. In explosives, the key reactions include

$$^6\text{Li} + n \rightarrow {}^4\text{He} + T + \text{energy}$$

$$D + T \rightarrow {}^4\text{He} + n + |rmenergy$$

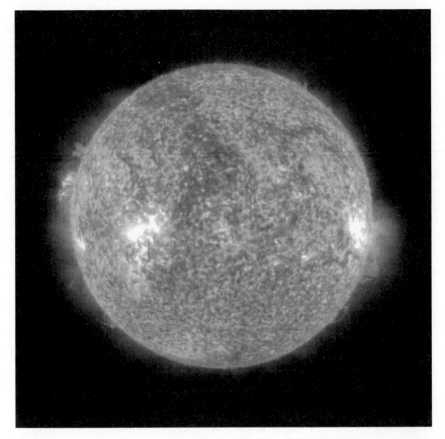

Fig. 6.4. Nature does fusion. The ultraviolet image of the Sun hints at the power of its energy source, even after being damped through 700,000 km of overlying material. Image obtained by the NASA/ESA Solar and Heliospheric Observatory (SOHO) on 6 May 2005, from Goddard Space Flight Center.

$$n + {}^{238}U \rightarrow \text{more neutrons}$$

Lithium can be chemically bonded to hydrogen, including the heavy versions deuterium and tritium, which helps keep the materials in place more easily than gaseous forms. Lithium hydride is a readily storable solid, which has the bonus feature for bomb-makers of fissioning to yield additional tritium when exposed to high-energy neutrons.

Returning to the interior of a star like the Sun, the reaction chain picks up quickly after the formation of deuterium. In a hot bath of protons, the deuterium nucleus will almost instantly fuse with another proton to form light helium (^{3}He). After another long wait, this nucleus will join with another of the same species to yield a "normal" helium-4 nucleus plus two loose protons. The net accounting in the whole process takes four protons and generates a nucleus of helium-4 plus transformation of 0.7% of the initial mass of the protons into energy. It was relativity that showed the connection between the "mass deficit" found on comparing hydrogen and helium and the release of nuclear energy. The phenomena of radioactivity were discovered before relativity, but it took the famed expression $E = mc^2$ to show more fundamentally where the energy comes from.

These reactions require extraordinarily high temperature and density to proceed, taking place only in a small core with about 10% of the Sun's mass and a thousandth of its volume, where the temperature reaches 14 million K. The tug-of-war between their energy input and the star's gravity defines the properties of stars and their evolution. The key was pointed out by Henry Norris Russell in 1919 – nuclear reactions act as thermostats. Without a central energy source, gravity would contract the star, driving up the central temperature and pressure. However, the rate of nuclear reactions rises very rapidly with temperature – some reactions can be approximated with powers as high as 100.[2] This leads to a stable balance between energy generation and overlying pressure of the star; any increase in pressure drives up the rate of nuclear energy production, heating the core and counteracting the weight of the star's outer layers. Conversely, if the nuclear reaction rate climbs above the balance point, the star's core expands and cools until equilibrium is reached.

As fusion proceeds in a star's core, its composition changes, and so do the rates of various nuclear reactions. As helium builds up, hydrogen nuclei encounter each other less frequently, so the core must be hotter to maintain the same outward force against gravity. This is because the pressure of the outer layers of the star exercises the nuclear thermostat – if the energy release from the core drops below the level needed to maintain the star's size, the core shrinks and heats until the energy production establishes equilibrium once again. While a star is in this phase, known as the main sequence from its location in the diagnostic Hertzsprung–Russell diagram, it gradually brightens, nearly doubling its luminosity as the hydrogen is gradually exhausted in the core.

All this action takes place deep inside the star. Even for the Sun, we cannot see any radiation coming from the core, through 700,000 kilometers of overlying matter. But there is a way we can see to the Sun's core, and check up on this theoretically-derived scheme. The $p - p$ reaction shown above is one of several in the solar interior that involves the *weak nuclear force*. This force can, among other things, transform a proton to a neutron plus antielectron (that is, a positron). When it does so, an additional particle is produced – a neutrino. Neutrinos interact with ordinary matter only very weakly (there is a reason this is the weak force). They stream directly out of the solar interior, and could penetrate a couple of light-years of lead before being significantly absorbed. This would make them wonderful vehicles to probe events at the Sun's core, except that they are equally unlikely to stop and register themselves in our detectors.

---

[2] This feature was exploited by at least one astrophysics professor, who was concerned that students were pushing too many buttons and thinking to little. He assigned problems in stellar reaction rates that would cause numerical overflow in calculators or computers unless the students took the trouble to convert the starting values into logarithmic form.

Detections of solar neutrinos date to 1967, many requiring heroic efforts to operate detectors massive enough to actually detect anything. These installations have to be located at the bottoms of mines, deep lakebeds, or under mountains, to reduce interference from other kinds of particles which might otherwise overwhelm the detection systems. The pioneering work was by Ray Davis, then at Brookhaven National Laboratory, who used a 450,000-liter tank of perchloroethylene situated 1,478 meters down in the Homestake Mine of South Dakota. Early calculations showed that the small fraction of highest-energy neutrinos from the Sun, arising in a side reaction involving beryllium and boron nuclei, would interact with 4–10 chlorine atoms per day, transforming them into a radioactive isotope of argon, which could then be counted. The results of Davis' experiment were consistent over its three decades of operation – the Sun emits neutrinos, but at no more than half the level predicted by our understanding of its structure and nuclear reactions.

For years, we were stuck with this "solar neutrino problem". Physicists and solar astronomers implicitly pointed fingers as to who did not properly understand their subject. Was the problem in our model of the solar interior, or in our understanding of neutrinos? After several large international projects, the answer is in – advantage, astrophysicists. Neutrinos come in three flavors (partnered with the electron, the muon familiar from cosmic-ray showers, and a heavier particle known as the τ). Neutrinos can cycle between these flavors as they propagate, with a rate depending on energy. The key observation came from neutrinos produced not by the Sun, but by cosmic rays interacting with matter in the upper atmosphere, which a neutrino detector can see at all distances from a hundred kilometers or so to the full diameter of the Earth. The Super Kamiokande detector, 1.6 kilometers deep under Mt. Ikena Yama in Japan, used Cerenkov radiation to detect energetic neutrinos, as they caused flashes in 50 million liters of extremely pure water. This kind of detection gives directional information, so the Kamiokande system yielded an image of the Sun (albeit with a resolution of about $10°$), one unique in being produced equally by data taken during the day and night. The oscillation of neutrinos between their various flavors implies that neutrinos must have mass, although it is so small that its value is still poorly known. It also brings the rate of detected solar neutrinos into good agreement with expectations from the fusion processes that should be happening in the solar core. This is one astrophysical problem which could be solved.

But the time has come to leave this byway – this is quantum mechanics, and you started off to read about relativity.

6.1 Nuclear alchemy – written in the stars

As stars shine, they convert lighter elements into heavier ones. Heavier ones yet can be produced in the violent deaths of stars, filling the periodic table. The results of this process are written in the code of starlight, as we work from the spectra of stars to the chemistry of their surfaces. Oddly enough, until the very end of its life, a star's surface does not betray the fusion taking place within. The vertical mixing needed to move newly minted atoms to the surface does not take place until the final stage of its existence as a red-giant (except for red dwarfs, which we will ignore for this purpose). When we measure the abundances of various elements in the spectra of stars (Fig. 6.5), they are generally the same mix that the star was born with. Thus we see the chemical heritage of each star, reaching back through previous generations of stars to the first gigantic beacons born in the early Universe.

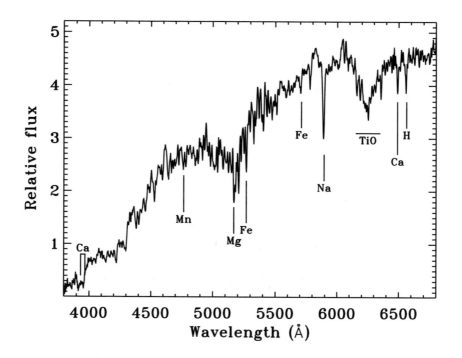

Fig. 6.5. Spectrum of the red giant γ Draconis, the star that first showed us aberration due to the Earth's orbital motion (Chapter 2). A few of the absorption features from gas in its atmosphere are marked, showing the numerous processed, heavy elements produced in previous generations of stars. Only in still later phases of their lives do most stars mix their own new products to the surface for our inspection. Most of the unmarked lines are additional features of iron, the most abundant element so deep into the periodic table. Red giants are cool enough for the existence of a few molecules, such as titanium oxide. Data from the 1-meter Anna Nickel telescope of Lick Observatory.

Patterns in these abundances tell us about the history not only of element creation, but the dynamics of stars and the formation of our Galaxy. We see correlations between the total abundance of heavy elements and such quantities as estimated ages of stars, their motions and distribution in the galaxy, and kinds of clusters they can form. These heavy elements, which is to say the ones beyond helium, are a distinct minority constituent in most stars. In the Sun, they add to only about 1% of the atoms. Furthermore, they vary more or less in lockstep with each other's abundance, which has led astronomers to refer simply to a star's metallicity.[3]

The details of stellar composition can tell us what kinds of stars came before. The relative amounts of carbon, nitrogen, oxygen, silicon, iron, and so on carry clues to what exact processes formed them, and in what stellar environments. Many of these pathways were worked out in a classic project by Geoff and Margaret Burbidge, Willy Fowler, and Fred Hoyle, published in 1957 and known universally by the authors' initials as B^2FH. Their initial interest was sparked by wondering whether *all* elements beyond hydrogen could be consistently produced within stars – a conclusion that would have profound cosmological implications. This work brought together four of the most prominent astrophysicists in a generation.

Geoff and Margaret Burbidge have been towering presences on the astronomical scene for almost half a century now. Geoff Burbidge made theoretical contributions in areas from the origin of elements to cosmology, often exercising his penchant for loudly questioning anything that smelled of orthodoxy. His long tenure at the University of California, San Diego, was punctuated by a term as director of Kitt Peak National Observatory, where he brought a distinct style centered on support for the community of observers.[4]

Margaret Burbidge could hardly have provided a greater contrast in style to Geoff. Her interests lay anchored in observation, particularly of quasars. She headed the team building one of the initial spectrographs for *Hubble*, and directed UCSD's Center for Astrophysics and Space Science. She served a term as President of the American Association for the Advancement of Science. Early experiences with chauvinist treatment left her not only as an advocate not for women in the sciences, but with an egalitarian spirit[5] at odds with the vestiges of British class consciousness that indirectly contributed much to science in the US. When she was named to a term as Director of the Royal Greenwich Observatory in 1972, that position was separated from that of Astronomer Royal for the first time in two centuries. Then, at Mt. Wilson, Margaret was informed that women could not be offered a research fellowship, with mention made of inadequate toilet facilities on the mountain. (As it happened, Geoff took the fellowship, and Margaret happened to appear every time he was scheduled on a telescope).

---

[3] In this, we have simplified the ancient Greek periodic table of the elements – earth, air, fire, and water. Astronomers often make do with the trio of hydrogen, helium, and metals.

[4] A minor accomplishment was hiring me at Kitt Peak for my first postdoctoral position.

[5] One of the things graduate students learned during her frequent observing trips to Mt. Hamilton was that she treated everyone in the same way and as an equal – visiting professors, graduate students, cooks and mechanics.

Fred Hoyle (1915–2001, eventually Sir Fred), perhaps even more than Geoff Burbidge, spent a career questioning authority even after becoming one in many fields. He championed the steady-state cosmology and several variants, which gave him a natural interest in whether all elements beyond hydrogen might not be star-forged. Under his direction at its founding, the Institute of Astronomy in Cambridge was quite distinctly the Institute of Theoretical Astronomy. His work on stellar evolution as well as nucleosynthesis earned him international reputation, and his willingness to defend unpopular ideas with intelligence and conviction kept him as one of the most prominent astronomers of the century.

Willy Fowler (1911–1995) continued to work in nuclear astrophysics at Caltech, receiving the 1983 Nobel Prize in physics for this work[6]. He led a group performing laboratory measurements of nuclear reactions that were important in astrophysics, and explored the implications of these results for tracing cosmic history through the isotopic makeup of stars.

Elements heavier than iron cannot be formed, except in incidental quantities, in normal stars – this is one of the reasons we believe that the Sun must have been a product of three or four generations of starbirth in the Milky Way. These elements require far hotter environments, such as might be found in exploding supernovae. Many of the heaviest elements have a pattern of occurrence which matches what results from almost any material flooded with vast numbers of neutrons (known as the r-process, for rapid addition of neutrons). We know this pattern not only from theoretical and laboratory studies of reaction and decay properties, but from samples left behind after what cosmochemist A. G. W. Cameron has delicately described as "terrestrial r-process experiments" carried out between 1945 and 1963 (again, Fig. 6.3). If we were to examine a site of the cosmic r-process very soon after it occurred, we would see very heavy elements, the ones beyond uranium which are so short-lived that we know them only from laboratory synthesis. And indeed, some supernovae have shown the characteristic decay rates of being powered at times by decay not only of lighter radioactive nuclei such as ^{56}Ni, but californium. Californium (element 96, four steps beyond uranium) is not found naturally on Earth, with its longest-lived isotope having a half-life of only 800 years. (Lighter radionuclides of titanium and aluminum can be unambiguously measured by the specific gamma-ray energies they give off during decay, and are seen in supernova ejecta. These have half-lives from 60 to 700,000 years.)

The origin of most heavy elements fits a fairly simple picture. Sunlike stars, fusing hydrogen in their cores, convert this to helium, and small amounts of beryllium and boron. These add to significant amounts of these light species believed to have been left over from the early Universe, as the expansion shut down collisions between protons and neutrons, and neutrons not safely bound into nuclei by the strong nuclear force decayed into proton–electron pairs.

As stars evolve to become red giants, the *triple-α process* can meld three helium nuclei to make one of carbon. One might at first wonder whether this should not proceed through two-nucleus collisions, which must be enormously

---

[6] Again, the prize committee attracted criticism for the perceived slight to Hoyle.

more common than three-way matchups. The bookkeeping would work out to have two helium-4 nuclei making one of beryllium-8. However, a glance at a periodic table of the elements shows that there is no such thing as beryllium-8. This combination of protons and neutrons is highly unstable and falls apart within about 10^{-16} seconds. As Fred Hoyle noted, another step in the process of nucleosynthesis would be possible if the energy levels of carbon are such that there is a high probability that the arrival of a third helium nucleus within this 10^{-16} second yielded an excited carbon nucleus (excited into a nonzero energy state to make the mass and energy books balance). Indeed, such energy levels do exist, a fact which Hoyle found profound, as it marked an important step in producing the raw material for our own existence. Most red giants shine by this helium-carbon transformation. Residual protons (hydrogen nuclei) can interact with the carbon to make nitrogen and oxygen, accounting for a trio of elements that we find especially interesting.

Low-mass stars never get beyond this level of nuclear development. They end their energy-producing lives as white dwarfs after gently puffing away their outer layers, often as planetary nebulae. We can even recognize the signature of carbon at the surfaces of some white dwarfs.

More massive stars can compress their cores to run hotter, slamming together nuclei faster and faster, and eventually yielding yet heavier elements. Their processing can run up through production of neon, magnesium, silicon, and so on. The most abundant products step through the table of elements in steps of two, both because these nuclei are often easiest to form and because remnant helium brings two protons with each nucleus. For some massive stars, when they become red supergiants, their nuclear structure resembles an onion. The outer layer is still pristine, mostly hydrogen and helium. Below this is a shell fusing hydrogen, with a helium-to-carbon shell inside that, and so on inward.

The most massive stars will carry this process to its end – iron. An iron core spells the death of a star, unfolding within milliseconds. With no net energy output from the core, gravity has its way and collapses the star. The result is a Type II supernova, which produces yet heavier elements as the shock wave races outward through the star, and then sprays much of the yield of these heavy elements out into the interstellar medium, where it will be incorporated into subsequent generations of stars. In some cases, curious creatures such as ourselves will isolate these heavy elements and construct intricate machines from them, some for the purpose of probing their own origins.

Some elements still pose mysteries, in that these stellar processes cannot produce nearly as much of them as we see around us. Gold and fluorine are too abundant to fit the basic pattern, and more exotic origins have been suggested to explain their abundances. Colliding white dwarfs and neutron stars have been brought forward. What an amazing circumstance it would be for the gold we prize to have been forged in such rare and violent events!

Further reading:

The story of Rutherford and Lord Kelvin is told, among other places, in David Lindley's *Degrees Kelvin: A Tale of Genius, Invention, and Tragedy* (Joseph Henry Press, 2004)

John Bahcall's article at http://nobelprize.org/physics/articles/fusion/index.html traces the early history of our understanding of the role of fusion.

Information on nuclear reactions in weapons is from the very useful *U.S. Nuclear Weapons: The Secret History* by Chuck Hansen (Aerofax: Arlington, Texas and Orion Books, New York) and *Nuclear Weapons Databook Volume I. U.S. Nuclear Forces and Capabilities*, by Thomas B. Cochran, William M. Arkin, and Milton M. Hoenig, (Ballinger Publishing Corporation, subsidiary of Harper and Row, Cambridge MA, 1984).

7 Extreme spacetime bending: black holes

Black holes are the objects *par excellence* of relativity theory. They mark the most extreme examples of spacetime curvature, time dilation, gravitational deflection of light, and gravitational redshift. What happens within a black hole remains inaccessible, shut off from our Universe as completely as the parallel worlds of science fiction.[1]

Astronomers have found compelling evidence not only that black holes exist, but that they are common. Some result from the collapse of massive stars, and others have been built up over cosmic time to equal a billion Suns in mass. These supermassive black holes appear to play a fundamental role in the evolution of galaxies, perhaps even introducing the feedback which sets the sizes of galaxies. The brilliant beacons of quasars are a side effect of the growth of these galactic anchors.

7.1 The river of spacetime

The effect of mass in general relativity is usually described as a bending of spacetime. This shows why, for example, light is deflected near masses (Chapter 4) and even why objects stay in orbits, closed paths in the curved background. But it is not so intuitively obvious why objects initially at rest should fall into regions of greater curvature, much less be unable to avoid falling into black holes. Andrew Hamilton and Jason Lisle at the University of Colorado have recently followed up a mathematical analysis first presented in 1921 by Allvar Gullstrand and Paul Painlevé, in what they term the "river" model for spacetime. By taking different coordinate choices than in the traditional "curvature" view (and after all, one point of relativity is that coordinate choices are arbitrary), they find a mathematically equivalent but much more intuitive description. Objects move, following the dynamics of special relativity, through a spacetime which is flat, but moves toward masses. In the extreme case of a black hole, this inward motion of the spacetime itself exceeds the speed of light, so an object cannot help being dragged in (being restricted to move slower than light against the current).

Hamilton and Lisle go on to show that the equivalence is complete with predictions of the standard approach, and much more intuitive to understand. Spacetime around a rotating black hole flows radially inward, but develops a

---

[1] And some interpretations of both cosmology and quantum physics.

twisting motion at each point – otherwise known as the dragging of inertial frames, now being tested in orbit (Chapter 9). It also becomes easy to see where the time dilation of light from material close to the event horizon, or falling through it, arises. Light has to swim upstream against the inflowing current of spacetime, so it will take progressively longer from farther in toward the central mass. Eventually, when the current speed exceeds c, it cannot escape at all. (And the rules of special relativity still apply: as with the rapid expansion in the epoch of inflation postulated by some cosmologists, it is acceptable for this current to exceed c so long as no information is transferred).

7.2 Black holes in the mind's eye

The concept that we now know as "black hole"[2] has a long intellectual pedigree. Between 1795 and 1799, John Michell in England and Pierre-Simon Laplace in France considered the issue of escape velocity. In Newtonian dynamics, the escape velocity from a certain point within an object's gravitational field depends on the object's mass and distance from its center. If one imagines shrinking a massive object, so that all its mass still contributes to its gravity for points very close to the center, one eventually reaches such an extreme density that the escape velocity exceeds the speed of light. Such an object would obviously be invisible from its own light. The notion was revived when solutions of Einstein's equations around very dense objects likewise showed the existence of regions which could never be seen from outside. The "radius" of such a region is twice what the Newtonian calculation showed.

Einstein's 1915 publication of general relativity set theorists off to find exact solutions of his equations under various conditions. Within months, Karl Schwarzschild published a solution for the case of an isolated point mass, and defining what we now call the Schwarzschild radius – the region from within which no signal can escape. He was unable to pursue this work, dying shortly thereafter of a disease contracted while he was in the trenches during the First World War. His solution included the region within the Schwarzschild radius, but it was not clear for many years whether that part of the theory was valid. Eventually it could be shown that an apparent interruption (a singularity) in physical quantities and coordinates crossing this radius was dependent on the choice of coordinate system, and that the implications for the interior were mathematically valid. Valid, and strange. The motion of particles inside the black hole sees at least the calculated roles of space and time interchange, as they approach ever closer to the central singularity.

Some physicists have found relativity's prediction of singularities to be the most disquieting aspect of black-hole theory. The equations say that the intensity of gravity (spacetime curvature) increases without limit, and that the size of the central mass will decrease without limit, toward zero. Physical laws are

---

[2] A memorable name coined by John Wheeler in 1969, replacing "collapsar" or some even less evocative description.

often undefined at zero radius; consider the classical Newtonian gravitational law, or the electrostatic force between charges. In both cases, looking at zero radius means dividing by zero, giving a mathematically indeterminate result. One can picture violations of cause-and-effect at such a point. Since we do not see such violations propagating through the Universe, physicists proposed a Cosmic Censorship Hypothesis: if singularities do exist, they are never seen naked, but always decently clothed in an event horizon so that these indeterminacies have no role in the outside Universe. The event horizon borders the region Schwarzschild had identified, from which no signal can escape. When we loosely speak of a black hole's size, this is usually what we mean.

Since Einstein had demonstrated how mass and spacetime are connected, a black hole becomes more complicated if it is rotating. More precisely, a black hole will retain the angular momentum of all the material which has formed it. This spin is reflected in the shape of the event horizon, and in its influence on surrounding spacetime. In 1963, Roy Kerr worked out solutions of Einstein's equations in this case, and found that the event horizon becomes surrounded by a region in which matter cannot remain stationary, and can in fact be flung outward with more energy than it had to begin with. This region is known as the *ergosphere*, whence energy can be extracted, and has been considered as part of the energy source for quasars and other active galaxies.

More complicated solutions for Einstein's equations were eventually found upon adding the third (and final) property of Einsteinian black holes – electrical charge. Chandrasekhar became fascinated by this ultimate simplicity of black holes. In general relativity, a black hole's properties are completely specified by three values: mass, spin, and charge. No other macroscopic object is so simple. Once a black hole has been formed, it has no memory of the detailed properties of the material that made it up. This is usually summarized by saying that "black holes have no hair".

This has become a contentious issue with the ambitious attempts to force a marriage between general relativity and quantum physics. Since we find that quantum processes becoming increasingly important on smaller and smaller scales, it seems clear that we will not know whether there is a singularity in black holes without this combination. Furthermore, the quantum fluctuations taking place constantly in "empty" space can have a dramatic impact on black holes. Hawking showed that these fluctuations will lead to the random emission of particles from just above the event horizon (*Hawking radiation*), which will lead to black holes of small mass evaporating over time. A black hole with the mass of an asteroid formed in the early Universe would just now be vanishing in a burst of gamma rays (although a burst quite different in detail from the ones we actually see). Larger black holes have not begun to evaporate, not only because their Hawking radiation is vastly weaker, but because they gain mass at a greater rate than they lose it, from the overall radiation entering their event horizons. The evaporation comes about because the energy to produce Hawking radiation has to come from somewhere, and that somewhere is the total mass of the black hole.

Hawking went on to ask what happens to information lost in a black hole. "Information" in physics is a measure of the order in a system; it may be quantified by the entropy, which measures how many different ways the constituents could be arranged (or ordered) while matching the observed condition of the system. If black holes truly have no hair, they are potent entropy generators, losing information at such a rate that Hawking long held that they must spawn new "baby Universes" to contain the information that vanishes. In a widely-publicized talk in 2004, he announced that he no longer considers this likely; Hawking radiation at the end of the black hole's existence will be nonrandom and contain some information about what went in. This information would not be particularly detailed, but enough to remove our suspicions about channelling the lost information into new Universes.

In 1939, J. Robert Oppenheimer (thereafter of Manhattan Project fame) and collaborator H. Snyder examined the mathematics of what were then called frozen stars. They found that under certain circumstances, massive stars might collapse into such unseen objects. Theoretical study of black holes largely languished until the 1960s, when John Archibald Wheeler began a renaissance in the application of relativity to such extreme situations (and gave us the name still widely used in the English-speaking world, although direct translation into French or Russian has undesirable connotations; Russian astrophysicists long preferred "frozen stars").

From our external vantage point, in a first approximation we would look at such an object and see ... nothing. In more detail, we would see things having no place in the classical physics of Newton. Light from distant stars would be wrapped around the black hole. Not only would we see multiple images or rings from background stars, but even stars directly behind us (and if we could look carefully, ourselves) would be lensed multiple times by rays of light wrapped once, twice, and more around the event horizon. If we dropped something into the black hole, we would see its light progressively redshifted, dimming constantly, and its motion slow down continually as it approached the event horizon. Relativistic time dilation would increase without limit as it approached this surface of no return. In its own reference frame, it would continue accelerating right through (and for a massive enough black hole, tidal forces would be weak enough that instruments could survive the passage). But from outside, we would never actually see anything fall through the event horizon. This leads to a paradox – if nothing ever crosses the event horizon in our reference frame, how could we ever see a black hole actually exist? It would indeed be a paradox if black holes could exist in the reference frame of material falling in but not an external reference frame. The solution lies in the fact that the spacetime signature of a collapsed mass depends less and less on the details of its distribution as one looks farther away from it. So, from a distance, the signature of a black hole matches more and more exactly that of mass piled up around the notional location of the event horizon. This is why it makes sense to talk of encountering black holes, rather than masses of material which we will never actually see forming them.[3]

Could such bizarre objects be found in the real Universe? Finding out posed formidable challenges to a science founded on the study of radiation from distant objects, rather than its absence. We might be able to see processes around black holes, and their interaction with visible matter. The search for black holes has proceeded mostly by process of elimination – we observe evidence for a large amount of mass in such a small volume that it could remain invisible in no other form we can think of than being a black hole. Such evidence might come, in principle, from several techniques. If we are extraordinarily lucky, we might see a black hole pass so nearly in front of a background star to see a signature of gravitational lensing. We might see stars moving in orbits around something massive and unseen. Or we might look for material being whipped closely around a massive invisible object, losing energy and being accreted across a one-way barrier. The most obvious and effective way to find back holes has been described by Dan Weedman, former head of NASA Astrophysics, as "listen for the screams of things falling in". When subjected to the enormous velocities and temperatures of the environment of a black hole, even atoms emit high-energy shrieks that we can detect from across the cosmos. These arrive in the form of X-rays from binary systems and the nuclei of galaxies.

In probing the cosmos, we can ask whether there are ways to produce black holes by natural processes, without the intervention of superintelligent aliens, quantum pressure anvils, or magic pixie dust. Our understanding of astrophysics has suggested that three kinds of black holes can be formed. So-called stellar (or stellar-mass) black holes would result from the collapse of a massive star when it can no longer generate nuclear energy to counteract its own gravity. This might be accompanied by a supernova expelling most of the star's outer layers, or simply collapse to invisibility without such an explosion. In the dense environment of a galactic nucleus, black holes would be able to grow by accreting surrounding gas; if massive enough, such a black hole could ingest whole stars, eventually becoming a supermassive black hole of perhaps billions of solar masses. Such an object would have an event horizon the size of our Solar System. At the opposite extreme, it may have been possible for chaotic conditions in the very early Universe to compress small amounts of material so strongly as to form tiny, primordial black holes – the mass of an asteroid in the volume of an atom. Of these, we have good evidence for stellar and supermassive black holes, and can at least tell how many primordial black holes we do not see out there.

Black holes have many bizarre properties, worked out by no one quite as deeply as Stephen Hawking. Much of the fascination with black holes centers on the interpretation of somewhat abstruse mathematical results. Does time actually stop? Can a black hole lead to its opposite number, a "white hole"

---

[3] This circumstance was used to chilling effect by Poul Anderson in his short story "Kyrie", in which a human is in telepathic contact with an alien who has been lost into a black hole. She is doomed to always "hear" its final thoughts, assuming telepathy to follow gravitational time dilation.

in some distant part of the Universe? Can black holes in principle be used to generate energy or travel in time? But astronomers' interests are much more modest. They are happy to call something a black hole if it meets the criteria of being so massive, invisible, and small, that nothing else we know of will fit. The really bizarre spacetime and entropy properties are quite secondary for most of us.

7.3 Black-hole hunting I: one star at a time

X-ray astronomy began with serendipity and has remained in its friendly grip ever since. Traditional astrophysics, circa 1960, knew that we could measure X-rays from the Sun, adding to less direct evidence that its corona was intensely hot. However, any other stars' coronae would be ten trillion times fainter, and not worth chasing with the available detectors. However, we would also expect to detect X-rays from the Sun as they were scattered from the Moon, which would provide hints as to the makeup of the lunar surface. Such ideas could only be tested with space experiments, since X-rays are effectively absorbed by the atmosphere at a height of 40–100 km (for various X-ray energies). At midnight on June 12, 1962, Riccardo Giacconi and colleagues at American Science and Engineering launched what amounted to a set of Geiger counters (Fig. 7.1) on an Air Force Aerobee rocket, to look for this scattered X-radiation from the Moon. In its five minutes above the atmosphere, the experiment failed to detect the Moon, but instead picked up a far more distant, and interesting, source in the same part of the sky. As Scorpius X-1 revealed itself, astronomy, and our visions of a placid and serene Universe, would never be the same. (And X-rays from the Moon? They were finally detected by a far more sensitive satellite payload in 1971.) Giacconi's team also encountered one of the most enduring mysteries of the X-ray Universe in their initial data – a faint glow pervading the entire sky, the X-ray background. This phenomenon will also have a role to play as we follow the hunt to unveil black holes and their history. [4]

Numerous sounding-rocket flights carried increasingly sensitive and sophisticated X-ray detectors aloft over the following years, turning up more X-ray sources and pinning down the locations of such strong sources as Sco X-1. These studies attained a new degree of completeness with the launch, on December 12, 1970, of the *Uhuru* satellite, designed to survey the entire sky for X-ray sources. As Riccardo Giacconi points out, each orbit of this satellite matched the cumulative observation time for X-ray astronomy throughout the 1960s. *Uhuru* was launched from an Italian platform off the coast of Kenya, on Kenya's independence day – hence the name, Swahili for "Freedom" (a considerable improvement

---

[4] Giacconi went on to apply his technical and managerial talents to a series of X-ray satellites, and eventually to the monumental task of building up the Space Telescope Science Institute so as to be able to adequately support *Hubble's* mission. He was awarded the 2002 Nobel Prize in physics, for his pioneering role in X-ray astronomy. Much of the history of X-ray astronomy is summarized in his Nobel lecture, available at http://nobelprize.org/physics/laureates/2002/giacconi-lecture.pdf.

Fig. 7.1. The detector payload used in the Aerobee sounding-rocket flight which discovered Scorpius X-1 in 1962. (Courtesy of R. Giacconi.)

over the original designation of Small Astronomy Satellite 1 or SAS-1). *Uhuru's* sky survey provided our first real view of the richness of the high-energy sky. Over the course of 27 months, it found over more than three hundred X-ray sources (Fig. 7.2), sampling all the major classes of object important in the X-ray sky: clusters of galaxies active galactic nuclei, flickering emission from compact remnants of stars, and the remnants of supernova explosions. Both the stellar X-ray sources (uniformly found to occur in binary systems) and active galaxies have proven important in the hunt for black holes.

As so often happens when new spectral windows are opened, optical identifications of the X-ray "stars" proved crucial to understanding them. Astronomers already knew, from the X-ray detections alone, several crucial facts about some stellar X-ray sources. Rapid flickering showed that the X-rays come from tiny regions, no more than a few light-seconds (and eventually light-milliseconds, which is to say hundreds of kilometers), in size. Furthermore, some showed periodic vanishing or diminution of the X-rays, just as we see when the ordinary light from binary stars is eclipsed as one stars passes in front of the other. When we can track them down in visible light, all the bright stellar radio sources in our galaxy turn out to come from binary star systems, of very particular types. Invariably, we can see one of the stars, but the other shows no sign in the optical spectrum. Analyzing the type and motion of the star we do see, we find that the invisible one has to be very compact and massive – either a neutron star or a black hole. When the X-rays show eclipses, a match of orbital periods between the X-ray source and potentially identical binary star system makes the identification ironclad.

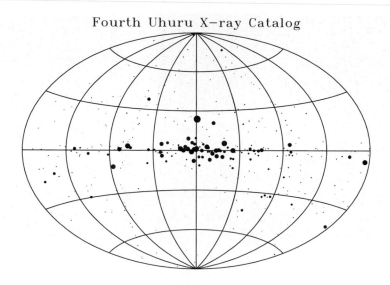

Fourth Uhuru X−ray Catalog

Fig. 7.2. The catalog of X-ray sources detected by the *Uhuru* satellite, shown in galactic coordinates. Larger symbols indicate stronger X-ray emission, on a logarithmic scale since the brightest ones are thousands of times stronger than the dimmest ones shown. Scorpius X-1 appears just above center, outshining anything else on the map. Sources near the equator lie toward the plane of our Galaxy, and are generally associated with accreting neutron stars or black holes. Objects away from the galactic plane are mostly distant clusters of galaxies, or active galaxies and quasars. Data from the NASA/Goddard Space Flight Center High-Energy Astrophysics Archival Research Center (HEASARC).

The search for *bona fide* black holes among these binary X-ray sources concentrated on those that could be best observed, and where the orbital plane is viewed in a direction that reduces the error in the unseen object's mass. The nature of the visible star is important – we must be able to infer its mass reliably from the spectrum, so it should be of a familiar kind with no sign that the black hole has, for example, removed enough of its mass to give it an unusual combination of mass, temperature, and luminosity. Attempts to prove the case for the first stellar-mass black hole proceeded slowly and deliberately, at least from the viewpoint of the community as a whole. Some of the individual astronomers probably thought everyone else was much too skeptical. Still, caution is wise in looking for a first documented example of something so far beyond previous observation; the first case needed to be as unimpeachable as the limits of observation would allow.

Gradually the cases strengthened through the 1970s for several of these X-ray binaries to contain black holes. The best-known and brightest is Cygnus X-1. Its optical counterpart is a blue supergiant star (Fig. 7.3), whose regular variations in Doppler shift trace a 5.6-day orbit in a binary with an invisible object. Based on the spectrum of the visible star, which is a hot supergiant with a mass estimated at 15 times the Sun's mass. Together with the orbital motion

Fig. 7.3. The massive black-hole binary system Cygnus X-1. At left is the soft X-ray view from the ROSAT observatory. At right, a visible-light image shows the supergiant companion star at center, and the brightest star shown. Each image shows a region of sky 10 arcminutes square. (Left: ROSAT data archive retrieved from the NASA Goddard Space Flight Center. Right: University of Alabama 0.4-meter telescope.)

of this star, that implies that the unseen companion, around which the X-rays originate, must have about 10 solar masses. This is comfortably above the mass limit for a neutron star, so the case for a black hole is strong.

Our census of potential black holes has widened dramatically with improvements in X-ray detection, each step yielding a richer X-ray sky. The (appropriately named) *Einstein* observatory, launched in 1978, employed focussing X-ray telescopes rather than the previous approach of exposing an X-ray detector to a region of sky limited by absorbing baffles (to a zone a degree or so across). This resulted in dramatic improvements in not only angular discrimination, but sensitivity, and led to a spectacularly successful series of missions by NASA, ESA, and the Japanese Institute of Space and Astronautical Science: Exosat, Ginga, ROSAT, ASCA, culminating in the flagship *Chandra* and *XMM/Newton* missions. These have grown in capability, size and complexity through the last three decades (Fig. 7.4), and X-ray work has become a mainstream part of astronomy.

Astronomers have identified additional black-hole candidates in binary systems, limited not by the X-ray data but by the observability of the companion star. Both in the Milky Way and in the Large Magellanic Cloud, there are now many X-ray sources with well-observed companions stars for which the unseen component has too much mass to be anything but a black hole. There are also many X-ray binaries in which the unseen component has too little mass to be a black hole. These neutron-star binaries can, it now seems, also be recognized by their X-ray behavior. If the X-rays brighten because of an influx of material to the surrounding disk, there may be a second flash as some of this material

Fig. 7.4. This montage shows the development of both technology and size of X-ray observatories. Across the top: *Uhuru* in assembly, showing its small size; *Einstein* during checkout before its 1978 launch, which marked the first use of a focussing X-ray telescope using grazing-incidence mirrors; and the European Space Agency's EXOSAT, seen in a mockup on display at a 1988 exhibition. Bottom: rollout of the *Chandra* X-ray observatory in preparation for its 1999 launch. (*Einstein* and *Chandra* photographs courtesy of Northrop Grumman Technology; *Uhuru* image courtesy of the Smithsonian Astrophysical Observatory.)

finds its way onto the surface of the neutron star (at half the speed of light). If it falls into the black hole, there would be no such flash – it vanishes. Confirmation of this idea would provide a powerful way to get a complete census of black holes (at least with close companion stars), and thus potentially work out just what kinds of stars collapse in this way. At this point, we are not even sure whether the formation of a stellar-mass black holes happens during a supernova explosion, or the star simply winks out.

7.4 Black-hole hunting II: the hidden center of the Milky Way

The center of a galaxy is a special environment. Here we find the greatest concentration of stars, and energy loss during collisions of interstellar gas leads to a growing amount of this material working its way inward over the life of a galaxy. We see all manner of fireworks in the centers of other galaxies (as set out in the following section), so it is natural to ask whether something similar happens at the center the Milky Way, a mere 28,000 light-years away, where we could examine it in great detail. This quest is hopeless for traditional optical astronomy, because the dust in several spiral arms blocks our view (Fig. 7.5). However, there are other bands in the spectrum which can penetrate this absorption with ease, leaving us ample opportunity to peer inward in our own Galaxy.

For the first three decades of infrared astronomy, mapping a piece of the sky had to be done the hard way – by sweeping the telescope back and forth so that a single small detector could look at each point. Not only was this slow, but changes in the atmosphere could impose stripes on the resulting image unless reference measurements were taken many times a second. This made it common for infrared telescopes to have small secondary mirrors which could be wobbled, or chopped, very rapidly between two positions, while the electronics took care of knowing what signal arrived when the mirror was in each position. As these detectors improved in sensitivity and wavelength range, the Galactic Center was a favorite target, not least because we knew from radio surveys exactly where to look, and each new advance had revealed new details.

First, we find the central concentration of stars, which should occur in any massive galaxy. This would be much like a globular cluster but larger, richer, and including the results of much more chemical processing within stars (Chapter 6). Simple extrapolation from the parts of our Galaxy we can see easily, and analogy with the Andromeda galaxy, suggested that the stars at the core should all be old, with their light dominated by the red giants. But something extra has been happening in our Galactic Center. We see giant regions of starbirth, both by seeing the brightest stars that they have formed and by the signature of the gas they illuminate. Nothing like this happens in the center of Andromeda. Some of the richest and brightest star clusters in the Galaxy are found here, within a few tens of light-years of the center. Here we also find a contender for the most massive star in the Milky Way, the "Pistol Star", named for the shape of the nebula created as its intense wind collides with surrounding gas.

Fig. 7.5. This wide-angle photograph shows the richness of the sky as we look toward the center of the galaxy, and the frustrating role of dust obscuration as we look into the center itself. Labels are given to prominent nebulae and star clusters, and constellation lines are shown for orientation. The slanting line marks the Galactic equator. Notice the crowding of dark clouds rich in dust toward this plane, and especially toward the Galactic Center itself. Picture taken, using a 35mm camera, from Cerro Tololo, Chile.

The whole Galactic Center region was picked up as a collective radio source by Grote Reber in the 1930s. Here too, each advance in our observational abilities has revealed unexpected phenomena, some still understood only in the sketchiest fashion. Here we find threads of radio-emitting material, twisted filaments tracing a large-scale magnetic field, a small spiral pattern which looks as if material is being pulled inward, and a tiny central source (most of these being seen in Fig. 7.6). The central source is denoted Sgr A* (as in "Sagittarius A-star" – do not look for a footnote), and is tiny for its power. Very-long-baseline measurements show it to be smaller than the size of Saturn's orbit. It did not escape notice that material around a massive black hole could radiate from a region this small. This radio source is also highly variable – a feature it shares with the vastly stronger central sources of quasars and other active galaxies.

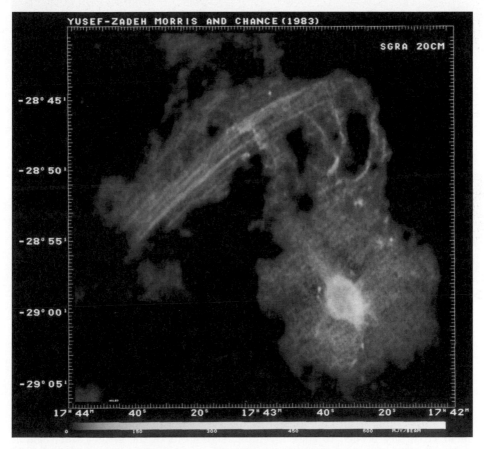

Fig. 7.6. The Galactic Center at a radio wavelength of 20 centimeters, zooming in to the inner few light-years. This image, from the Very Large Array, shows the twisted filaments, central source Sagittarius A with Sgr A* at its heart, and some of the bright spiral structure around it. (Image by F. Yusef-Zadeh, M.R Morris, and D.R. Chance; courtesy of NRAO/AUI.)

X-rays, at least at higher energies, can penetrate the gas between here and the Galactic Center, providing a rich and surprising view of conditions there (Fig. 7.7). We see enormous bubbles of gas, heated to 2 million K, testifying to explosive events within the last few thousand years. Michael Muno of UCLA recently led a team finding an unusual number of X-ray binary systems within a few light-years of the center, suggesting that the Galactic Center contains an extraordinary clustering of perhaps 10,000 black holes and neutron stars. And there is the central X-ray source, precisely aligned with Sgr A*, flaring in brightness every few days.

Fig. 7.7. X-ray mosaic of the Galactic Center, obtained with the *Chandra* X-ray observatory. The Galactic Center is surrounded by clouds of gas at several million K, as well as a concentration of stellar X-ray sources. At the very center sits the variable source which may represent accretion of gas into the central massive black hole. (NASA/University of Massachusetts/D.Wang *et al.*)

The case for some kind of compact central mass in the Milky Way started to grow through infrared spectral observations, which showed dramatic gas motions in the inner few light-years. These included work published by a Berkeley–Caltech team in 1979, with ground-based data, and 1984 work by a larger team including some of the same people, now using the 0.9-m telescope of NASA's Kuiper Airborne Observatory (Fig. 7.8). The Earth's atmosphere, particularly its water vapor, absorbs radiation very efficiently in the deep infrared (thereby being the main player in our greenhouse effect). Making matters worse, it radiates furiously at these same wavelengths, and so ground-based astronomy at these wavelengths is like trying to pick out stars in the daytime through a pair of sunglasses while someone is shining a searchlight on them. Infrared astronomy made remarkable strides before spacecraft, all the more remarkable for the technical virtuosity needed to coax a reliable signal out of the strong and variable atmospheric noise. While the best results are obtained above the atmosphere, aircraft can fly high enough to get above most of the water vapor and allow observations which are impossible from the ground. Frank Low of the University of Arizona (and later a key figure in the IRAS project) pioneered airborne astronomy by fitting a 30-cm telescope into a Lear Jet and equipping it for infrared use. Eventually, NASA constructed the Kuiper Airborne Observatory – an aircraft fitted with a 0.9-meter telescope and capable of taking it anywhere on Earth. Unlike satellites, airborne telescopes can be quickly repaired, refitted, and reflown, so that any observation which can be done from the air can usually be done faster and cheaper (if not always "better"). The Kuiper Observatory was so successful that it was eventually retired to make funds available for development of a successor, the US–German Stratospheric Observatory for Infrared Astronomy (SOFIA). This is a long-range Boeing 747 carrying a short-focus 2.5-meter telescope, now undergoing system tests.

Fig. 7.8. NASA's Kuiper Airborne Observatory, a C-141A transport plane modified to carry a gyrostabilized 0.9-m telescope which looked out of an opening just ahead of the port wing. Part of the spoiler door covering the opening can be seen outlined just ahead of the wing. The telescope was carried high enough to escape most of the water-vapor absorption which hampers far-infrared astronomy, allowing measurements without the expense of a satellite and the ability to change and reconfigure instrumentation. The aircraft is shown in its hangar at the NASA Ames Research Center.

Infrared spectra near the Galactic Center showed gas in rapid motion, at angles strongly inclined to the more placid disk and spiral arms. The simplest analysis suggested motions in the gravitational field of a central mass (perhaps a black hole) of several million solar masses. However, this kind of analysis was quite rightly regarded as suspect, as it is in other galaxies. Huge clouds of interstellar gas can be shoved around by forces other than gravity. Shock waves from supernovae and interstellar magnetic fields can give these clouds motions which have nothing to do with nearby masses. The definitive test would be in measuring the dynamics of stars. Stars are so dense that once formed, their orbits are essentially fixed to respond only to gravity. If the orbits of stars also told us there were millions of solar masses in something tiny and unseen at the Galactic Center, that would be worth believing.

Here again, this evidence trickled in as our technological capabilities advanced. Infrared astronomy from the Earth's surface has been revolutionized by two new technologies, both of which began (at least in the US) in the classified realm. First to come into astronomical use were infrared array cameras. Much of the original technical development here was originally driven by the military needs for night-vision devices on the battlefield, and space sensors capable of detecting distant missile plumes. As they became available for civilian use, astronomers began to push their detection performance yet further. No longer limited to a single pixel scanning the skies, infrared astronomers could now take images much as optical astronomers were accustomed to. Arrays reached 1,024, then 4,096, and by now a million pixels. The sky could be mapped – very deeply and at high resolution, surprising us with unexpected results. There is hardly a field of astronomy which has not felt the impact of infrared arrays, from high-redshift galaxies to star clusters to the outer Solar System. Looking toward the Galactic Center, it was now possible to carry out a serious census of the stellar population, at wavelengths that penetrated much of the intervening dust. These detectors also allowed spectroscopic measurements of newly faint objects, practiced almost as we would work in visible light. Doppler shifts, temperatures, and chemical compositions can now be derived for stars too cool, or too heavily reddened by dust, to measure in visible light.

The second technology is adaptive optics. Astronomers at least since the time of Newton have fought the blurring effects of atmospheric turbulence. These can be reduced, but not eliminated, by selecting observatory sites with the smoothest possible airflow, and then by designing telescope structures so that we do not produce our own turbulence immediately in front of the telescope. Still, every telescope with an aperture larger than 1 meter and many smaller ones find their performance limited not by their optical quality, or the ultimate limit set by the wave nature of light, but by our atmosphere – in astronomical parlance, the *seeing*. Until recently astronomers tended to view this as part of the weather – we all seem to complain incessantly without actually doing anything about it. To be sure, there had been attempts to break the atmospheric limit. Robert Leighton obtained remarkable photographs of Mars from Mt. Wilson using electronic sensors to follow the atmospherically-induced image motion, a technique now in regular use by numerous backyard astronomers with webcams. Beginning in 1970, Antoine Labeyrie introduced astronomers to the art of coaxing information from atmospheric blur patterns frozen by very short exposures (known as speckle interferometry). This technique proved fruitful in measuring binary stars normally blurred together by atmospheric seeing, but frustrating for more complicated images.

In the 1970s, astronomers and national-security planners alike realized the potential advantages of removing atmospheric blurring on the fly, using a flexible "rubber mirror" to change the optical properties of the system many times a second, so as to compensate for the changing properties of the atmosphere.

Tracking these changes would require some kind of reference source – normally a bright star. Lacking such a star, one could create an artificial one, by shining a narrow beam from a laser into high layers of the atmosphere and using it to measure the distortion. Working at the University of Hawaii, Laird Thompson led a project to test such laser guide stars as early as 1978. But such sporadic efforts within the astronomical community were to be rapidly overtaken by military efforts, which put some of the relevant technology out of the public eye for years.

By the first launch of the Space Shuttle *Columbia* in 1981, it has been widely rumored that a couple of Air Force sites imaged its protective tiles with adaptive-optics cameras. Within ten years, declassification of these systems' operating details was providing astronomers with new possibilities for all their existing tools. Whole systems were handed over, operating principles were being discussed at public meetings (in some cases showing that the astronomers had done a farther-reaching analysis of the theory), and observing time was being awarded for astronomical projects on the Air Force 3.5-meter telescope at Haleakala, Maui. Specifically astronomical adaptive-optics systems were under development in Europe, Canada, and the US. By the mid-1990s we were seeing results from these systems, which soon realized their promise for infrared work. The engineering problem is more tractable in the infrared; sensors can watch the visible light and make corrections, sending the infrared on to a camera or spectrograph, and the image quality is better and not quite so rapidly changeable in the infrared. With the new generation of 8- and 10-meter telescopes, adaptive optics let them realize the angular resolution inherent in their apertures, and deliver finer detail than even *Hubble's* infrared camera NICMOS delivered.

Looking at the Galactic Center, the changeover from speckle interferometry to adaptive optics yielded one of the most spectacular scientific results yet obtained from the 10-meter Keck telescopes atop Mauna Kea (Fig. 7.9). A team led by Andrea Ghez at UCLA was able not only to measure individual giant and supergiant stars near the Galactic Center, but so track their motions as they orbit the central mass. These orbits have periods as short as 15 years, so that a few years' data have pinned down their orbital shapes and sizes (Fig. 7.10). On top of this, they have been able to use the adaptive-optics system to isolate the light of individual stars and measure their changing Doppler shifts, so that we know their orbits in three dimensions. Virtually identical results came from Reinhard Genzel's group in Germany, using the European Southern Observatory's Very Large Telescope (an array of 8.2-meter instruments in Chile, and now the best-instrumented observatory on the planet). From the orbits of these stars, tracked with precision across significant arcs of their paths, we see that there is a dark central mass in the Milky Way, of slightly more than 3 million solar masses. This mass must be smaller than the closest approach of any of these stars, a radius of 45 Astronomical Units (6 light-hours, slightly more than the average distance of Pluto in our own Solar System). Based solely on the motions of these stars, the evidence for a massive black hole at the Galactic Center is compelling.

Identification of a black hole at the center of the Milky Way, and measurement of its mass, opens a whole new set of questions. Why is not it quite as massive

Fig. 7.9. Moonlight emphasizes the white domes of the twin 10-meter Keck telescopes atop Hawaii's Mauna Kea. Fifteen years after the first of these segmented-mirror instruments went into operation, they remain the world's largest optical telescopes, and have become an integral part of the advance of astronomy. Their construction and operation was overseen by a consortium of Caltech and the University of California, with NASA support. The lights of the Big Island's Kohala coast are at right, and the 8.2-meter Subaru telescope of the National Astronomical Observatory of Japan is partially seen at left.

as we would expect from other galaxies like the Milky Way? How did young massive stars get into orbits so close to it, when they clearly could not have formed there simply because the black hole's tidal forces would have shredded protostellar gas clouds? Finally, it is very quiet compared to many other black holes in galactic nuclei; is its quiescence only a temporary stage?

7.5 Black-hole hunting III: the black hearts of galaxies

As the flickering Universe of binary X-ray sources was being revealed, observations across the spectrum were showing that violent and energetic events take place on a titanic scale in the nuclei of galaxies. These were revealed in several guises, as we discovered active galactic nuclei.

Seyfert galaxies were grouped as a class by Carl Seyfert in 1943, based on inspection of the modest number of galaxy spectra available on photographic plates at that point. E.A. Fath had recorded the strong emission lines of NGC 1068 as early as 1908, but had no standard of comparison to assess its spectrum with anything other than relief that it had strong emission peaks which were easily recorded. Seyfert noted a small subset of galaxies whose spectra showed

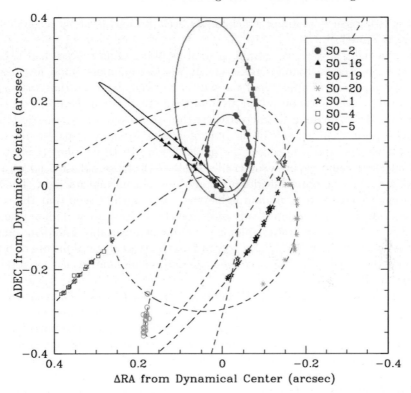

Fig. 7.10. The orbits of stars near the Galactic Center, as traced by infrared adaptive-optics imaging with the 10-meter Keck telescopes. Symbols mark observed positions of luminous stars (catalogued as S0 followed by a running number), with their fitted orbits superimposed. Solid curves show the orbits which constrain the central mass most directly. These orbits indicate a central mass, less than a light-day in extent, which is spatially coincident with the compact radio source at the Galactic Center and totals over 3 million solar masses. The entire diagram represents an area that lies within the blur pattern of a single star on most traditional astronomical images from ground-based telescopes. Image provided by Andrea Ghez (UCLA).

strong and unusually broad emission lines from their nuclei. These emission lines have ratios different from what we see in star-forming regions such as the Orion Nebula, indicating a more exotic power source than even large numbers of hot stars. Later studies showed that these nuclei could be broadly classified into two types, depending on whether all the emission lines showed similar Doppler widths, or differences showing a higher-density inner region as well. Eventually, Seyfert nuclei were recognized as strong X-ray sources, and as variable through most of the electromagnetic spectrum. These nuclei are too small for telescopes to resolve – even the eye of *Hubble*. The lessons of Seyfert nuclei can be summarized as telling us that very rapid gas motions take place in regions too small for us to distinguish – motions which might well be anchored by a very strong gravitational effect.

Radio galaxies were recognized as soon as many radio sources had been observed with even early interferometers. It was astronomers' good fortune that this common kind of object had the only nontrivial structure that was easy to recognize by eye on the output of an interferometer – two separate peaks rather than one. Their defining characteristic is the appearance of symmetric lobes of radio emission on either side of the optical galaxy. In many cases, we detect jets of emitting material connecting the galaxy's core to these lobes (Chapter 2). The nuclei of radio galaxies sometimes show Seyfert-like emission spectra, but sometimes show no optical evidence of anything but normal starlight. The radio lobes may stretch several million light-years, adding the lesson that the central objects can eject material in steady directions for at least a few million years and possibly rather longer. Radio galaxies are almost invariably found in elliptical galaxies or their disturbed kin, in contrast to the typical spiral nature of Seyfert galaxies. This may mean that the galaxy environment controls the expression of nuclear activity.

Quasars were originally found as another product of radio astronomy, even though most of them are not strong radio sources. The first quasar identifications were of strong radio sources with no obvious galaxy counterpart, for which precise positions led optical astronomers to undistinguished-looking starlike images. Their spectra showed redshifts of unprecedented magnitude. To appear so bright, quasars would have to be extraordinarily luminous, far outshining any kind of normal galaxy. Indeed, it was an ill-defined part of the definition that a quasar had to look like a star on typical photographs, and it has required substantial efforts on the ground and in space to peek through the quasars' glare for good views of the surrounding galaxies. Quasar spectra share the main features of Seyfert galaxies' spectra; to a large extent, a Seyfert nucleus is like a quasar with the volume turned down. Beyond their extraordinary energy output, the new lesson of quasars was that this energy is released in a tiny volume. The light output of quasars varies erratically, on times from days to months. Quite generally, this means that the light source cannot be much larger than the distance light travels in these times. In itself, these variations show that we are not dealing with any normal products of stellar processes, even on the most intense scale.

These kinds of active nuclei share a family resemblance, and must all be facets of a single phenomenon. High-energy emission in X-rays is common for all, and the spectral features of glowing gas are similar among the three. Many quasars show double lobed and radio jets as seen in radio galaxies, while quasars and Seyfert nuclei have similar variability processes.

By the mid-1970s, the leading contender to explain the central engines was a supermassive black hole, with the accompanying fireworks produced by matter trapped in orbits close to it and eventually falling in. The high orbital velocities provided a natural way to heat material to X-ray temperatures, and the expected disk of circling matter provided a geometric platform to launch jets of relativistic matter (Chapter 2. To some extent, this scheme won popularity because of the lack of any plausible alternative. Still, the mass and size scales we had to account for meant that, as Martin Rees summarized the possibilities in a 1978 review, anything that small and massive either is a black hole already, or will become one soon.

Peter Young played a crucial role in collecting some of the first compelling dynamical evidence for black holes in galactic nuclei. As a beginning graduate student, he had developed an interest in how massive black holes would affect the surrounding stars. In particular, he calculated the distribution of stars expected around such a black hole. Factoring in their orbits, he found that galaxies with a massive black hole at the center would have strongly peaked concentration of starlight within the inner tens of light-years, in contrast to galaxies without them, where the central peak would be much smoother and less pronounced. Furthermore, the motions of these stars would be much more rapid in the presence of the black hole, simply because the orbital velocities at a given radius would have to be higher in the presence of extra mass.

A chance to test these predictions came in mid-1977, when Young was a postdoctoral researcher working with Wallace Sargent of Caltech. The first astronomical charge-coupled devices (CCDs) were becoming available, and they attached one to the 5-meter Hale telescope at Palomar for observations of galaxies that were strong candidates for hosting massive black holes. They selected radio galaxies, which show signs of a long history of activity and occur in elliptical galaxies in which the stars' orbits populate all directions about the core. M87, in the Virgo cluster, was particularly intriguing, with its unusually energetic jet (Fig. 2.11) and exceptional optical luminosity. The precision of the CCD images, over a wider brightness range than previous detectors, allowed them to measure the light profiles, and hence star distributions, in these galaxies with the accuracy to tell whether central cusps were present. In hindsight, their choices were indeed clever; most other galaxies they could have observed have cusps too small to be clearly distinguished without the resolution later delivered by *Hubble*.

Sargent and Young also investigated the motions of stars in the inner regions of M87, in what became the prototype of similar investigations to the present day. They used a very sensitive detector based on a TV camera coupled to an image intensifier, developed by Alec Boksenberg in the UK and transported to the world's largest telescopes by "Boksenberg's flying circus". They were able to measure the spread in Doppler shifts of the stars point by point along a slit crossing the core, showing that these internal motions increase toward the center in just the way expected for motion around a central mass with not millions, but several billion solar masses. A similar mass was implied by the distribution of the stars, which made quite an impact when the two results were published in back-to-back papers. Both these results were based on the motions of stars, which is important since glowing gas is easier to measure but can be affected by forces other than gravity. As with the center of our own galaxy, the case for a central mass is always more reliable when it is based on the motions of nearby stars.

Their conclusions were soon challenged, as scientists are wont to do. The issue was not their data, but technical issues of interpretation. How much do we have to know about the orbits of stars in a galactic nucleus if we are to make sense of the velocities we can measure? The measurements use Doppler shifts, and therefore tell us about the motions along the line of sight. Some theorists pointed out that it was possible (if not necessarily plausible) for most stars to be in very elongated orbits about the core, so that we would see stars in their innermost, fast-moving phase only while looking at the core – black hole or not. Still further analysis showed that there was a strict limit to how many stars could occupy such orbits without interacting to put most of them in better-behaved paths, but it would dramatically improve our certainty to have data going yet closer to the core of some of these galaxies.

The image quality of large ground-based telescopes started to improve in the 1980s, as sites such as Mauna Kea and La Palma were exploited, and more attention was paid to the role of airflow and thermal equilibrium in image quality. A leader in this direction was the Canada–France–Hawaii telescope on Mauna Kea. This capability persuaded new Hawaii faculty member John Kormendy to take a decade-long detour from his work on elliptical-galaxy structures to pursue black holes in galaxies, where the CFHT provided a competitive advantage. The Andromeda galaxy was an excellent starting point; in 1988, Kormendy presented evidence for a central black hole of 10 million solar masses at its core. This was remarkable not only for the strength of the evidence, better than in M87 simply because we could see stars so close to the core, but because the center of the Andromeda galaxy is as quiet in visible light, radio, and X-rays as that of any big spiral galaxy. If Andromeda hosted such a black hole, so could every luminous galaxy.

A few other cases were brought forward in the following few years. The Sombrero galaxy, a giant spiral in Virgo; NGC 3115, an edge-on galaxy with a disk but without spiral structure; and maybe M81, a bright spiral only 10 million light-years away on the ear of the Great Bear.[5] Hopes were high that the impending launch of *Hubble* would resolve this, like so many other questions.

The new instruments on *Hubble* could reach their full potential only after the December 1993 servicing mission corrected its optical aberrations. The first set of spectrographs were not very well-suited to galaxy dynamics, simply because they could measure only a single region of a galaxy at a time. The progress made with these instruments was mostly in observing disks of gas orbiting central masses. A small disk showed up in M87; careful modelling by Linda Dressler at the Space Telescope Science Institute strengthened the case made by Peter Young and colleagues 15 years earlier. This case was strengthened by the European instrument team which had built the Faint-Object Camera, used in its spectroscopic mode. But it was the next generation of *Hubble* instruments that opened the floodgates.

The second *Hubble* servicing mission was performed by the crew of the Space Shuttle *Discovery* in February 1997 (Fig. 7.11). In addition to various maintenance tasks, they installed two long-awaited instruments which represented

---

[5] Not that many other astronomers noticed, but I did those particular observations from La Palma in the best atmospheric conditions I had encountered, before or since.

substantial extensions of the telescope's capabilities. One was the near-infrared camera NICMOS; the other, the Space Telescope Imaging Spectrograph (STIS). STIS provided flexible spectroscopic capabilities throughout the ultraviolet and visible regimes, and could measure all the light falling along a slit. This was just the tool to measure the Doppler dispersions of stars throughout a galaxy, and finding telltale signatures of black holes well beyond the abilities of ground-based telescopes.

Fig. 7.11. The Space Shuttle *Discovery* rockets into the predawn sky on February 11, 1997. The exhaust plume from its solid-rocket boosters illuminates the orbiter's tail fin, while its glow outlines the steam cloud surrounding the launch pad. During this mission, the crew installed the Space Telescope Imaging Spectrograph (STIS), which was to provide crucial dynamical evidence of central black holes in many galaxies.

As anticipated, STIS and *Hubble* were poised to tell us the answer. And they did – over and over. Using stars rather than less reliable gas clouds, STIS data revealed black holes in the centers of galaxy after galaxy. Dozens of such measurements had been made by the time STIS was disabled by an electronic failure in August 2004.

As the *Hubble* results in stellar dynamics on galactic cores accumulated, a pattern emerged, sometimes known as the Magorrian relation (after the first author on the paper that most people noticed). The mass of the central black hole varies systematically with the amount of mass in the surrounding stars, particularly if we consider stars in the round central bulges of the galaxies. Similar relations occur whether we look at the total brightness of these stars, or at their collective Doppler-shift velocities. A typical galaxy has a nuclear black hole of about 1/200 the mass of the surrounding bulge stars. If anything, our own galaxy is somewhat of an underachiever in this, which may tell us something important about why we are here. This relation implies that the existence, and possibly growth, of this central black hole is a normal part of the galaxy's formation and history, rather than an optional feature uncoupled from the surrounding galaxy.

Active galaxies seem to be black holes in the process of growth, if the action we see is a side effect of matter being accreted. This allows us a two-way check on how many black holes have grown to what masses. In the Universe here and now, *Hubble* has given provided a census of black hole masses in galactic nuclei (leading to the Magorrian relation). And as to how they got that way, deep X-ray surveys should be telling us directly how many black holes were growing at what rates over cosmic time. In fact, the collective X-rays reaching us from quasars and other active nuclei throughout cosmic time is the major contributor to the all-sky X-ray background found by Riccardo Giacconi so many years ago. A major *Chandra* study led by Amy Barger found that most of the growth occurred fairly early in cosmic time, more than 8 billion years ago. This is an interesting epoch, being the time when we see that quasars were most numerous. This would all imply that most bright galaxies have gone through at least one brief quasar phase. Today, the great majority of these black holes are quiescent, revealed only through their effects on motions of surrounding stars, and flaring into brilliant activity only when a fresh supply of gas is provided by such disturbances as the gravity of a passing galaxy.

As reassuring as it may be to see this concordance between the statistics of black-hole growth and today's masses, there are surely things we still need to learn about how these black holes came into being.

The Sloan Digital Sky Survey has, as intended, unearthed (that verb should probably be "unskied") the most distant known quasars, beyond $z = 6$. Using estimates honed on spectra of more nearby and recent quasars, these have black-hole masses just as large – hundreds of millions of solar masses – despite being seen as they were when the Universe was no more than 850 million years old. That is none too many years to form a galaxy and stars which then feed one central black hole. The kind of starting black hole we know about – the remnant of a collapsed massive star, with perhaps 10 solar masses – may not be enough to start the process this quickly. There may have been a role for some special process in creating the seed black hole, which could grow much more rapidly if it began with 100, or 1,000, or 10,000 solar masses. One progenitor discussed in theoretical studies off and on for years has been a relativistic star cluster – one so compact that the internal motions become relativistic, and the gravitational redshift for light leaving the cluster becomes important. Under some conditions, simulated relativistic clusters collapse rapidly to form central black holes; but under other conditions they are stable for astronomically long times, so it is not obvious that this is an answer.

Another way to form 100-solar-mass black holes would be the collapse of one of the first–generation stars. All the stars around us today have significant amounts of the heavy elements, those fused in earlier generations of stars (albeit less significant amounts for older stars). There must have been a generation of stars without these elements, forming from the hydrogen and helium bequeathed by nucleosynthesis in the first few minutes of cosmic history. These stars would have been much hotter and more massive than we find today, because of this lack of heavy elements. They would leave behind, at least according to some calculations, massive black holes after titanic supernovae expelled huge masses of heavy elements to seed subsequent star formation and the Universe we know. These explosions represent our most promising way to find these "Population III" stars, and may be within range of the next generation of space infrared telescopes.

Further reading:

Stephen Hawking, *Black Holes and Baby Universes and other essays* (Bantam, 1994)
The uncovering of the black hole at the Milky Way's core is described by Fulvio Melia in *The Black Hole at the Center of the Milky Way* (Princeton, 2003).

8 The shape of Einstein's Universe

In presenting general relativity, Einstein considered the application of curved spacetime to very large volumes – that is, potentially to the Universe itself. In evaluating the outcome, he was guided by the notion that the Universe was essentially static. Much has since been made of the philosophical preconceptions that such a choice might betray (most notably not having to grapple with "why is this here?" if it has *always* been here), but to astronomers of the time, the idea of an eternal Cosmos might have seemed so natural as to be almost subliminal. While Newton had wrestled with the problem of the stability of the system of stars (as far as he knew, the entire Universe) on the basis of gravity producing groupings of stars more and more strongly with time, it was only well into the twentieth century that astronomy became an historical science. Empirical evidence and theoretical frameworks have since shown that stars and galaxies alike have histories, and undergo dramatic changes with cosmic time. Indeed, among the most pressing astrophysical questions which drive the development of new facilities now are the formation of galaxies, stars, and planets. While the question of the formation of the planets had been debated at least back to the time of Laplace and Kant, astronomers at the turn of the last century saw little evidence that much ever changes in the Universe. Stars could be seen to move, but about as many were approaching us as receding. We had our first glimpses of the rotation of the Galaxy, but again, the mean motions of stars balanced out.

It must have been awkward for Einstein when he found that the basic formulation of general relativity yielded a Universe that would be either in constant expansion or contraction. Since these obviously did not match reality, he added a term to the equations, Λ, the cosmological constant. This corresponded to some yet-unobserved repulsive agent which became important only on cosmic scales, eventually counterbalancing the effects of gravity in pulling stars and systems of stars together. The cosmological constant has had a checkered career, appearing and disappearing from discussion as new kinds of data are available and evaluated. Although such new data induced Einstein to conclude that Λ must be zero, recent observations of the distant Universe have convinced most astronomers that, although we have no good theory for why, Einstein seems to have been right the first time (as we see at the end of this chapter).

Other theoretical physicists and mathematicians soon began exploring the consequences of general relativity in a wide range of situations – an activity which continues today. Several noted, during the 1920s, that various solutions for Einstein's equations had radically different implications for cosmic history.

In particular, an expanding Universe was actually a simpler solution (requiring no extra term such as the cosmological constant). This was noted by Willem de Sitter in the Netherlands, the Abbe Georges Lemaître in Belgium, and, most notably, by Alexander Friedmann in the Soviet Union. Einstein seems to have had Lemaître's vocation in mind as he expressed his skepticism of this scheme: "I have not yet fallen into the hands of priests."

Meanwhile, astronomers in the increasingly powerful observatories of the American Southwest were revealing the Universe of galaxies, showing that, however vast our own Milky Way, it was only one of innumerable similar clouds of stars stretching away beyond the limits of our greatest telescopes.

While it is Edwin Hubble whose name is most associated with this leap, important data had already been contributed by Vesto Melvin Slipher, using the 60-cm (24-inch) refracting telescope (Fig. 8.1) at Lowell Observatory in Flagstaff, Arizona. Earlier attempts had been made to photograph the spectra of what were then known as spiral nebulae, but their light is so diffuse that no useful results had been obtained, beyond the fact that they could not be composed of excited gas; their spectra were continuous, rather than comprised solely of emission lines. Slipher experimented with various combinations of dispersing prisms and exposure times, some as long as 35 hours spread over several nights. In a series of papers from 1911 to 1918, he reported Doppler shifts for fifteen such nebulae, ranging as high as 1,100 km/s. Such large velocities were not found for stars or gaseous nebulae which were clearly part of our Galaxy – a strong hint that we were dealing with something new. Slipher's observations are particularly impressive to us today. Classical refracting telescopes, such as the Lowell instrument, are unfriendly for spectroscopy, requiring careful compensation for the change of focus with wavelength. The available photographic materials were very insensitive by modern standards, requiring long hours of exposure on multiple nights, with constant mechanical corrections to the telescope's tracking, to leave a faint trace on the spectrogram.

Slipher's argument for the importance of spiral nebulae, as "one of the important products of the forces of nature", had in view the large numbers of these objects which had been found by James Keeler using the 0.9-meter (36-inch) Crossley reflector at Lick Observatory[1]. Keeler published an influential collection of nebular photographs in 1918, in which he could already point to some of the regularities that would occupy astronomers for decades. For example, he showed sequences of similar spirals arranged by their angle of inclination as we view them showing that disks of absorbing material (that is, interstellar dust) occur in their midplanes. Keeler had found thousands by photographing relatively small areas of sky, of all sizes down to the limit of the photographs' definition.

---

[1] This same telescope, with revamped tube and mounting, was still in use seventy years later when I was investigating the variability of the gravitationally-lensed quasar 0957+561. The longevity of many astronomical instruments may variously be either good stewardship, or a problem for funding agencies.

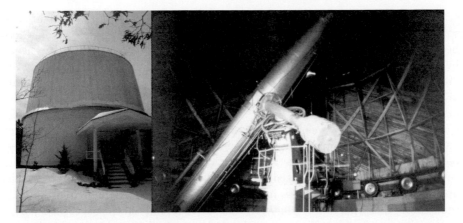

Fig. 8.1. The 24-inch (60-cm) refracting telescope of Lowell Observatory, overlooking Flagstaff, Arizona. This instrument, in the hands of V.M. Slipher, revealed the large shifts in galaxy spectra that were to be the earliest hint of cosmic expansion. The exterior picture shows the need for the snow passage around the door in the Flagstaff winter. For a brief time in the late 1890s, the dome floated on a water-filled track and was rotated by pulling ropes. It is said that on windy winter nights, when the water was salted to prevent it from freezing, the briny spray and creaking of the ropes made observers feel at sea. Salt corrosion in the circular trough put an end to this experiment.

The nature of these nebulae was clearly an important problem, and one that would likely yield only to the most powerful observational techniques. Two schools of thought dominated their study at this point, both with very respectable intellectual roots reaching into the eighteenth century. One view saw in the spirals the embodiment of Laplace's picture of the formation of planets, from material swirling around a nascent Sun and possible flung off from it. In this case, these would all be nearby, well within our Galaxy, and only the size of our Solar System. A grander view, harking back to Immanuel Kant and possibly slightly earlier to Thomas Wright in 1750, suggested that the vast system of stars we see as the Milky Way is only one of many such systems distributed through space over an indefinite extent. If so, the spiral nebulae must be very distant to appear so small and dim in our sky. Thus, an immediate way to distinguish these two notions would come from measuring the distances of spiral nebulae.

This was to be a delicate task. Astronomers often seem obsessed with distances, because they can be so hard to derive, and because so much of what we know about distant objects requires knowing its distance to transform what we measure into genuine physical properties. Parallax, triangulation using the Earth's orbit as a baseline, had succeeded in measuring the distances of nearby stars only in 1837. Associations of stars into clusters provided statistically useful, albeit individually less precise, values for more distant stars. This reach proved deep enough to reveal important patterns in the behavior of some kinds of stars – patterns which remain the gold standard in probing galaxy distances to this day.

Even before Hubble's work, there were already clear hints that some of these spiral nebulae were distant, independent systems of stars. Astronomers had long been aware of nova outbursts – stars which temporarily increased their brightness by such huge factors that the precursor had generally not even been seen, fading away over days or weeks. They did not yet distinguish between nova outbursts, in which a white dwarf in a binary system violently expels a layer of accreted material in a titanic thermonuclear explosion; and supernova explosions, which end a star's life and may completely destroy the star. Still, a nova (which we now know to have been a supernova, reaching almost naked-eye brightness) had been recorded near the center of the Andromeda Nebula in 1885, and work beginning with G.W. Ritchey in 1917 used the 60-inch telescope atop Mt. Wilson to turn up "novae" (most of which were in fact supernovae) in additional fainter nebulae, some found on inspection of photographic plates already in the observatory's collection. That same year Harlow Shapley (who was to have a long and influential tenure as director of the Harvard College Observatory) tabulated eleven such novae, and pointed out that the Andromeda object must have been extraordinarily bright in comparison to the novae we knew of in the Milky Way. Ritchey had been plain that the faintness of these novae implied nebular distances in the millions of light-years – clear support for the Kantian island-Universe picture.

Some of the ambiguities remaining in trying to understand the spiral nebulae were brought out in the 1920 "Great Debate" set out before the National Academy of Sciences, between H.D. Curtis of Lick Observatory and Harlow Shapley, then of the Mt. Wilson Observatory staff. This was not exactly a debate, but a pair of presentations from different viewpoints by scientists who had exchanged notes in order to answer one another's arguments more effectively. The subjects ranged widely, with the size of the Milky Way playing a prominent role. Shapley used observations of variable stars to argue that the Galaxy was much larger than hitherto estimated, so that the spiral nebulae would be quite small by comparison and relegated to substructures, perhaps gas collapsing to form stars. Curtis argued for a smaller Galaxy, but that the spirals must be independent systems. As we look back with the benefit of much better data, both were right in their major points. The Milky Way is nearly as large as Shapley held, if we trace the most distant stars belonging to it, but the spirals are in fact million of light-years away and quite comparable to the Milky Way, as Curtis proposed. Perhaps the greatest interest of this exchange, which makes it still worth reading, is in looking over the shoulders of scientists trying to make progress where the data remain ambiguous and sources of error plentiful. As Frank Shu wrote in his 1982 textbook *The Physical Universe*, "The Shapley–Curtis debate makes interesting reading even today. It is important, not only as a historical document, but also as a glimpse into the reasoning processes of eminent scientists engaged in a great controversy for which the evidence on both sides is fragmentary and partly faulty. This debate illustrates forcefully how tricky it is to pick one's way through the treacherous ground that characterizes research at the frontiers of science." Clearly, resolving this issue would require data of a kind not available to Shapley or Curtis. Such data arrived within a decade, from the work of Edwin Hubble at Mt. Wilson.

Fig. 8.2. The 60-inch (1.5-meter) telescope at Mt. Wilson, with which Ritchey found what we now know to have been supernovae, hinting that spiral nebulae are in fact distant galaxies comparable to the Milky Way. Its sturdy design appeared on film at the hands of Boris Karloff in the 1936 film *The Invisible Ray*, cinematically relocated to a Carpathian mountaintop and able to capture rays from the Andromeda galaxy in a driving thunderstorm.

8.1 Edwin Hubble and the galaxies

Edwin Powell Hubble arrived at the Mt. Wilson Observatory in 1919, already a veteran of World War I, study at Oxford, and careful observational work for his Chicago dissertation on nebulae. He was well prepared to make powerful use of the new 100-inch telescope which had been installed in 1917, and proceeded to make the new field of extragalactic astronomy virtually his own for two decades. His plan for probing the distances of galaxies built on a discovery made by Henrietta Leavitt at Harvard, analyzing photographs from the observatory's station in South Africa. Easily observed only from the southern hemisphere, our two brightest satellite galaxies, the Large and Small Magellanic Clouds (Fig. 8.3), furnished important laboratories for the study of stellar populations even before the idea of independent galaxies was firmly established. Their relatively small size in our sky means that all the stars, clusters, and nebulae in each Cloud are at roughly the same distance from us, allowing comparative studies without

Fig. 8.3. The Magellanic Clouds, Large at left and Small to the right, are the brightest of the Milky Way's satellite galaxies. Data for the Large Cloud, in particular, played a key role in showing how powerful a distance probe Cepheid variable stars could be. This 35-mm photograph, taken from Cerro Tololo, Chile, gives some impression of these objects' visual appearance from dark locations in the southern hemisphere.

the tedious (and sometimes impossible) task of independently estimating the distance of every star of interest.

A key result of these Magellanic Cloud studies pertained to the class of variable stars known as Cepheids, after the first-known member of the class (δ Cephei). These stars are supergiants near the end of their energy-producing lives, and exhibit combinations of temperature and luminosity which render them unstable. Instead of maintaining the relatively steady energy output of, for example, the Sun, these stars pulsate regularly about the mean, with the surface heating and cooling as the star expands and contracts. Just as a larger bell rings with a deeper tone, so an intrinsically brighter Cepheid takes longer to complete each pulsation. Leavitt found a remarkably neat relation between the brightness and period of Cepheids in the Large Magellanic Cloud; the period was linearly related to the star's average magnitude. Such a relation for Cepheids in the Milky Way had been obscured by their individually uncertain distances.

Armed with this relation, and even a crude calibration of the distance of nearby Cepheids, Hubble could seek similar stars in spiral nebulae. The most promising candidates were clearly those that appear largest and brightest in our sky, signposts of proximity when we know nothing else, and perhaps those that showed the clearest images of what could be individual stars. He obtained repeated series of photographs of likely regions in the Andromeda galaxy, the Pinwheel Galaxy M33 in Triangulum, and (first to yield to his technique) the irregular system NGC 6822 (Barnard's Galaxy) in Sagittarius. In 1925 and 1926,

Hubble detected multiple Cepheids in these galaxies, and demonstrated that the word "galaxies" has physical meaning.

Like many extragalactic astronomers after him, Hubble felt the call of cosmology, and devoted much effort to exploring ways in which measurements of galaxies could tell us about the structure of the yet greater Universe. Here, he was thinking of galaxies as test particles partaking of the overall history of spacetime, much as oceanographers might trace ocean currents using corks.[2] He worked with Caltech theoretical physicist Richard Tolman, who had already investigated the implications of various assumptions about the Universe for astronomical observations. By 1935, they set out ways to use the sizes and brightnesses of galaxies to sample cosmic history. These problems are complicated in one case by the fact that galaxies are fuzzy objects without well-defined edges, and in the other by our view combining galaxies with wide ranges of distances, luminosity, and redshift. Hubble maintained plans for a large project involving faint galaxy statistics for cosmology, particularly as the Palomar telescope approached completion. The essence of his method was to seek a change in the derived space density of galaxies with distance. Coupled with the travel time of light from such large distances, this would be the signature of changes in the expansion rate (what would later be known as the Hubble constant, although he was already seeking evidence that it was not a constant). However, by that time it had become clear that this approach would not yield reliable cosmological conclusions even based on Palomar photographs – something that was stressed to him by the other staff astronomers in eventually turning down his request for using huge slices of precious Palomar time in this project.

Hubble's examination of galaxy redshifts as a distance measure were more immediately fruitful. In this he was joined by an unlikely collaborator, Milton Humason. Humason's association with Mt. Wilson began as a mule driver during construction of the observatory. He exhibited a deft mechanical touch, and was eventually made an observer, directly responsible for operation of the telescope and many of the details of spectroscopic measurement. Allan Sandage later recalled that his personality made him an ideal foil to Hubble's aloof and distant manner. In 1931, Hubble and Humason jointly published an extension of Hubble's original redshift–distance relation, now combining Hubble's distance estimated with additional implicit estimates based on the observed brightness of typical galaxies (Fig. 8.4). These results were more convincing than Hubble's sparse points in his 1929 paper, and put the redshift–distance relation on a firm footing. Around this time Einstein visited Mt. Wilson (Fig. 8.5), and was photographed at the eyepiece of the 100-inch telescope. Discussion with Hubble seems to have convinced him that the Universe was indeed not static, and the cosmological constant disappeared from relativistic cosmology (at least until it was needed again).

Ever since this time, the linear relation we see in the local Universe, between a galaxy's distance and its redshift, has been called the Hubble law. Such a

---

[2] Or in one famous case in 1992, thousands of rubber ducks sent overboard in a Central Pacific storm. Few cosmologists believe their field to be quite this desperate.

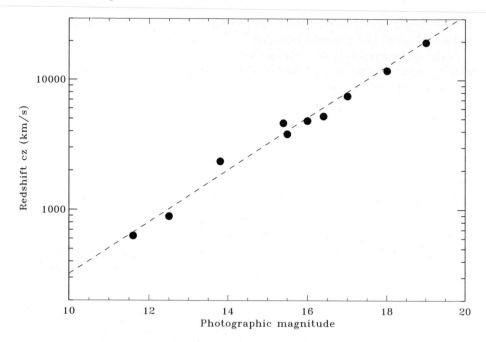

Fig. 8.4. The data available to Hubble and Humason in 1931, showing the case for a redshift–distance relation, later interpreted as an expanding Universe. Their values have been replotted here from an *Astrophysical Journal* paper. To extend the range of the available technology, they used several ploys. They considered that the typical luminosity of the galaxies they observed was the same everywhere, averaging the brightest members of galaxy groups, and added mean values for samples of non-cluster galaxies of similar apparent brightness. A linear relation on the plot (in which both axes are logarithmic) would imply a direct proportionality between redshift and distance – what has since been called the Hubble law.

linear relation implies a uniform expansion, in which everything was once in the same place at the same time. This cosmological redshift is subtly different in interpretation from the familiar Doppler shift. As radiation propagates through expanding spacetime, its wavelengths partake of this expansion. If we see a galaxy spectrum in which every feature is at a wavelength double that of the emitted "laboratory" values, this tells us that the entire scale of the Universe has doubled while the light was in transit. This makes a difference because, although the spectral redshift itself looks like a Doppler shift, the intensity of light reaching us is progressively less than for a pure Doppler shift. The familiar inverse-square law for light travelling through empty space breaks down. If cosmic expansions means that there is more space to fill after a certain travel time than ordinary Euclidean geometry predicts, there will be less light per square centimeter when it arrives, and the redshift factor also applied to the rate of arrival of photons as well as to their energies. The effect of the redshift is to stretch all the observed

Fig. 8.5. Einstein with a group including Edwin Hubble (left rear), during a visit to Mt. Wilson Observatory in 1931. This was the same year in which Hubble and Humason published newly extended evidence for the redshift–distance relation that was to become the cornerstone for our picture of an expanding Universe. (Copyright California Institute of Technology, used by permission.)

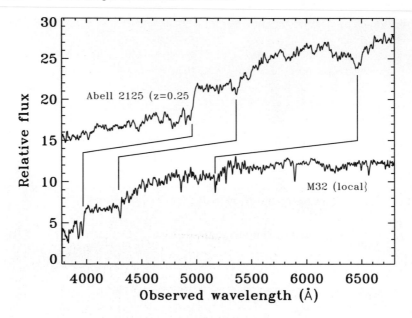

Fig. 8.6. Spectra of elliptical galaxies at low and significant redshifts. M32, at the bottom, is one of the companions of the Andromeda galaxy, and exhibits almost exactly zero redshift. The upper spectrum is the average of three bright elliptical galaxies in the cluster Abell 2125 at redshift $z = 0.25$, about 3 billion light-years distant. The lines mark the same spectral lines in both spectra, showing that the amount of shift increases linearly with wavelength. That is, the shift includes a stretch. The galaxies in Abell 2125 are more massive than M32, so their stars move faster, making their spectral lines broader. M32 data from the 1-meter Nickel telescope of Lick Observatory; Abell 2125 spectra from the 4-meter Mayall telescope of Kitt Peak National Observatory.

wavelengths by the same ratio $(1 + z)$. This is illustrated in Fig. 8.6, which compares the spectra of nearby and distant elliptical galaxies.

For nearly three decades after Hubble, the quest for the scale and history of the Universe virtually centered on one man – Allan Sandage, working at the Mt. Wilson and Palomar Observatories. His acumen with observational technique brought him to Hubble's attention, making Sandage his assistant during Hubble's final years. After Hubble's death in 1953, Sandage shouldered the burden and the vast opportunity of taking on Hubble's cosmological quest. From his position as the major single user of the Palomar 200-inch telescope (Fig. 8.7), he was uniquely positioned to make progress on the intricate path from stars to the Cosmos.[3] He undertook an ambitious, decades-long quest to uncover first the

---

[3] Which Sandage did, with enormous energy. A citation study of the astronomical literature found his work to be solely responsible for a statistically significant differ-

cosmic distance scale, in the form of the Hubble constant H_0, and then to describe the history and future of the Universe itself. In traditional relativistic cosmology, this quest was "a search for two numbers".[4] The first was the Hubble constant, relating the expansion rate of spacetime to the distance across which we evaluate it. In the local Universe, this is the ratio between the redshift of a galaxy (usually expressed as the velocity that would produce the same redshift from the Doppler mechanism, for convenience if not clarity) and its distance. Careful discussions refer to this as the Hubble ratio or the Hubble parameter, since it might change with cosmic time.

Fig. 8.7. The 200-inch (5.08-meter) Hale telescope on Palomar Mountain. This instrument reigned as the most powerful telescope on Earth for four decades, and played pivotal roles in such diverse fields as the cosmic distance scale, galaxy evolution, the discovery of quasars, an dthe search for black holes.

Such a possible change in the Hubble "constant" leads us to the second number, which describes the rate of cosmic deceleration. Since we know the Universe to contain matter, if not quite how much, its collective gravitational influence in any given part of the Universe (and by extension, the whole) would decelerate the overall expansion. One could picture, and describe mathematically, three cases. If the matter density is high, this deceleration would someday overcome the expansion completely, leading the Universe to collapse on itself in a "Big

---

ence in citation rates between papers authored by astronomers at universities and at private institutions.

[4] A phrase from the title of one of Sandage's articles.

Crunch", perhaps setting off another cycle of expansion, reversal, and contraction. The geometric properties of light travel in such a Universe are analogous to the geometry of lines on a closed surface such as a sphere, so this case is known as a closed Universe. At the other extreme, the density of matter might be so low that its deceleration would never reverse the expansion, leading to a Universe that would expand without limit, that could happen only once. Light propagation in such an open Universe would be analogous to geometry on the surface which was not only infinite but "outwardly curved", which can be visualized as a sort of saddle shape. And in between was the knife-edge balance of a flat Universe, where there was just enough matter to exactly counteract the expansion given infinite time. In such a Universe, the expansion would always be slowing but never completely stop. Light would propagate in a flat Universe following just the rules of Euclid for ordinary geometry. In this classical vision of cosmology, geometry and destiny were the same – the mass density of the Universe determined its expansion history and its fate.

Theorists describe the geometry of spacetime through a mathematical expression known as "the metric". The metric relates the distances one might travel (or light might travel) as measured along the various coordinates (such as the x and y of ordinary geometry) to the total distance travelled. The simplest metric relation, in the flat space of Euclid, carries Pythagoras' name. In curved space(time), the metric departs from this simplicity. Depending on the distribution of mass and how rapidly spacetime is expanding, cosmological distances might be either greater or less than this intuitive value, with the departures increasing as we look across more and more of the Universe.

8.2 Climbing the distance ladder

Sandage's attack on the Hubble constant used a multistep approach which has become enshrined in the jargon as the *distance ladder*. Starting with expertise on the theory of stellar life cycles gained in a postdoctoral year at Princeton, he laid out a methodical plan to work outward from the nearest stars to the most distant galaxies, using each set of data to calibrate the next. Almost as side effects, this quest laid important groundwork across much of contemporary astronomy – the ages of star clusters, physics of stellar pulsation, structures of galaxies, and statistics of galaxy clusters. The ladder begins with knowing the size of the Earth's orbit. This lets us triangulate to measure the distances of the nearest stars, as their apparent places shift against the distant background on a one-year cycle. This produces a small set of very accurate data on the luminosities as well as temperatures and motions of the most common kinds of stars. We can add to this set by finding stars of these kinds in more distant star clusters, knowing that all members of the cluster are at nearly the same distance. Eventually this leads us to distances of Cepheid variable stars, either in the Milky Way or our neighboring Magellanic Clouds.

Now we can jump the abyss to other galaxies, since Cepheids could be detected from Mt. Wilson and later the giant Palomar telescope throughout our

Local Group of galaxies. However, at such small distances of only a few million light-years, we learn nothing about the Hubble constant. The local gravitational motions, as the galaxies mill about in the overall gravity of the group, utterly dominate any Hubble component. In fact, the conventional interpretation of cosmic expansion is that it is an asymptotic phenomenon which does not apply to bound systems such as galaxy clusters. So we need to add rungs to the ladder, reaching so far away that these local motions are negligible compared to the cosmological redshift. Cepheids, observed with photographic plates from ground-based telescopes, could not be picked out so far away. They could, however, be used to calibrate other potential standard objects. The brightest stars in galaxies, the sizes of their largest nebulae, and the peak brightness of supernovae could all be pressed into service, later joined by the mean brightness of globular clusters and the brightness statistics of planetary nebulae. However, each of these steps was tedious, and its accuracy depended on all the preceding steps. Furthermore, there were issues of just how to correct the results for absorption by dust both within the Milky Way and in the other galaxies.

Eventually, distances would be needed for galaxies so far away that we could see none of their individual component stars. Here, the statistics of galaxies themselves had to be used. What was the typical luminosity of the spiral galaxies with the longest, best-defined arms? How about the brightest galaxies in clusters, or the third- or tenth-brightest? Were there regularities in the properties of quasars that might let us estimate their distances from observable properties?

With so many steps involved, it is easy to see why such alternative distance approaches as gravitational lensing and light echos have been attractive. It is also clear how appropriate was the name of the Hubble Space Telescope, given its ability to pick out Cepheid variable stars perhaps a hundred million light-years away and supernovaeindexsupernovae to nearly ten billion light-years.

By 1974, Sandage and long-time collaborator Gustav Tammann set out their results on this distance ladder, reported in a 21-year series of ten research papers entitled "Steps toward the Hubble Constant". They converged on a value close to 50 kilometers per second per megaparsec.[5] Beyond the distances to galaxies estimated using Hubble's linear relation, the Hubble constant tells us something fundamental about the Universe – its age.

In the simplest kind of expansion, we might imagine galaxies separating at the same rate forever, without deceleration. In this simple, easily computed, and physically uninteresting conception of the Universe, we can run the clock back to see when everything would have been in the same place at the same time. The answer is numerically the inverse of the Hubble constant. For a value of 50, the cosmic age would be about 20 billion years, enough to comfortably fit

---

[5] One megaparsec is 3.26 million light-years. The parsec has somehow become more established than the light-year in the professional literature, although both are rather arbitrarily referred to the Earth's orbit. The purist will note that dimensions of length appear in both numerator and denominator, so the actual units of the Hubble constant are inverse time. Once again, convenient numbers and practical usage trump strict usage; Sandage's value of 50 km/s per megaparsec translates to about 1.6×10^{-18} per second, which does not lend itself to vocal arguments quite as readily.

in the estimated ages of all the stars and star clusters examined in detail. More realistically, the age of the Universe would be somewhat less than this value, allowing for the expansion to have slowed over time as the gravitational effect of matter played a part. For a flat Universe with a so-called critical density of matter, the age would be 2/3 of the inverse of the Hubble constant – in this case, a bit over 13 billion years. That was not quite so comfortable, but would still work as long as galaxies took shape and began forming stars quickly. (This aspect has not changed with the subsequent changes in our value of the Hubble constant and introduction of a mysterious new force in cosmology.)

This exact issue – the history of cosmic expansion and its extrapolation into the future – was the second goal of Sandage's program of observational cosmology, albeit one that did not make such headway until there was more competition in the field. Rather than on the value of the Hubble "constant", this focussed on how that value might have changed over cosmic time. This would be reflected in distant objects appearing at somewhat different distances than would be expected from extrapolating the Hubble relation from the local Universe. One can even detect such changes without knowing the exact local value of the "constant", as long as the same kinds of object are compared at all distances. They must really be the same – even minor biases in how we select, say, galaxy clusters, with distance could swamp the cosmological effects. The signature of a change in the expansion rate would be curvature in the redshift–distance relation (in essence, the line of Fig. 8.4) when carried out to distances well beyond a billion light-years. In carrying out such measurements, a host of subtle corrections had to be made for the data in the near and far fields to be strictly comparable. As had been pointed out by Tolman soon after Hubble announced his redshift-distance relation, general relativity predicts several ways in which distant objects will dim in an expanding Universe. Since the radiation is redshifted as it reaches us, the energy of each photon will be lower by the factor $(1 + z)$. They will also be more spread out in time – another effect of the coordinate transformation between then and there, and here and now, by the same factor. Making things yet worse, since redshifting also stretches the entire spectrum in wavelength, our filters will pass a progressively narrow (and dimmer) piece of the galaxy's spectrum at higher redshifts. Put all together, the average surface brightness of a given galaxy will fall as the fourth power of $(1 + z)$. This makes working on high-redshift galaxies much more difficult than their sheer distance would suggest.

On top of these effects, we must know the spectrum of a distant galaxy to properly compare its brightness with local examples. Because of the redshift, the piece of a galaxy's spectrum that we measure with a particular filter was emitted at shorter and shorter wavelengths for higher redshifts, so we must know how to transform that into the intensity in the same piece of the emitted spectrum. Sometimes we can use a different filter that mimics light starting in our reference filter; sometimes we can measure the galaxy's spectrum directly across the wave;length region we need; and in other cases we have to model the galaxy's spectrum based on whatever data we have, and apply a calculated ratio. This is known as the K-correction, and by analogy the change in a galaxy's appearance when we look at it in optical, infrared, or ultraviolet bands has come to be known as the morphological K-correction.

Sandage's data on distant galaxies could not quite support a clear conclusion on cosmic deceleration, so he looked elsewhere for hints to cosmic history – most notably comparing the age of the Universe derived from the Hubble constant (known as the Hubble time) to the ages derived for the oldest stars, based on what we understood of stellar evolution. It was just as well, because major challenges had arisen to his conclusions about both the Hubble constant and its change over cosmic time.

The authority of Sandage's determination of the Hubble constant was questioned most insistently and vociferously by an energetic and opinionated Frenchman from Texas, Gerard de Vaucouleurs. Having worked with his wife and colleague Antoinette for years in cataloguing galaxies, he brought an encyclopedic knowledge of the field and a willingness to reconsider every step of the distance ladder. He was willing to include additional but less reliable distance indicators in his calculations, on the theory that they would not all be wrong in the same way, and could collectively serve as a check on the Cepheid distances. (This approach was also advocated by Sidney van den Bergh in Canada, albeit attracting less attention and acrimony.) By 1977, de Vaucouleurs was in print suggesting a much higher value for the Hubble constant. He favored 100 km/s per megaparsec, twice Sandage's best value; this put all distant galaxies half as far away and made the Universe only half the age Sandage had championed. Admittedly, this high value of the Hubble constant produced a cosmic timescale which was uncomfortably short for the ages of the oldest stars (returning to a perennial problem in cosmology, which does at least help us get off the wrong track occasionally). The difference between their distance scales, and final values for the expansion rate, did not arise in one straightforward place. Each looked at the whole set of distance indicators independently. The difference began before looking outside of our Galaxy, in how each corrected the observed brightness of distant objects for the absorption of light by dust in the Milky Way. Then each settled on a somewhat different distance for the Large Magellanic Cloud, which served as an anchor point for Cepheid variable stars. Then de Vaucouleurs pointed out that, since regions of the Universe with more galaxies presumably have the expansion held back to some degree by the gravity of the extra matter, a Hubble constant weighted solely by the number of galaxies involved would be biased to a low value. And so it went. Sandage and de Vaucouleurs traded barbs in print and at meeting for years. It is striking how personal and accusatory these issues became. Having the wrong value of the Hubble constant evidently was more than a mistake. It was a sin.

However, Allan Sandage and Gerard de Vaucouleurs did agree on one point about the Hubble constant. Anyone who simply took an average of their preferred values (50 and 100), and used 75 in calculations, perhaps hoping to reduce the risk of a factor-of-two error, was guilty of spineless fence-sitting, which was even worse than being bullheadedly wrong.[6]

At the height of this debate (50? 100? Long distances? Short?) a new generation of cosmological sleuths applied yet another distance indicator to land squarely in the middle of the debate. First, Brent TullyTully, Brent (working at the Observatoire de Marseille and shortly taking up a position at the University of Hawaii) and Rick Fisher (of the National Radio Astronomy Observatory location in Green Bank, West Virginia) followed clues scattered among earlier work by several authors to find that the Doppler width of cold hydrogen (observed at 21 centimeters) in spiral galaxies is very well correlated with galaxy luminosity. They showed this by selecting galaxies in which other unknowns were reduced. First, the galaxy had to be either at an otherwise well-known distance – Andromeda, or in a few nearby groups – or belong to a rich group so that an internal comparison made sense. Next, we had to view the spiral disk at a steep enough angle that the Doppler shift we see reflects almost the entire rotation velocity along the major axis, but not so steeply that dust absorption distorts the galaxy's overall brightness. Using this winnowed set, they found a surprisingly good correlation between the orbital speed of most of the gas and the luminosity of a galaxy's stars. In their first approach, they suggested a Hubble constant of 80 km/second per megaparsec. This so–called Tully-Fisher relation has been a cornerstone of galaxy distances ever since, eventually joined by the analogous Faber–Jackson relation for elliptical galaxies (in which we have to measure the stars' overall Doppler shifts, there being no gas to see at 21 centimeters). This relation made spiral galaxies themselves somewhat like Cepheid variables as potential distance indicators. A quantity which we measured independently of distance (the 21-cm Doppler width) was well correlated with something whose measurement is dependent on distance (the galaxy's overall brightness), so we could tell what distance makes the correlation work for any particular galaxy. In a way, the Tully–Fisher relation is an application of an important fact about galaxies. Big galaxies are big, and small galaxies are small, and this dictates so much of what we see that it obscures many more subtle effects.

The Tully–Fisher relation could be exploited not only to measure galaxy distances and thus calibrate the Hubble constant in yet another way, but could be used to map motions in the Universe. If we were to look at galaxies in, say, the nearby Virgo cluster, and find them to all be more distant than their redshifts indicate, we might conclude that some local motion between ourselves and Virgo is compromising our measurement of the Hubble "constant" within this region. This is just what happened when the trio[7] of Marc Aaronson, John Huchra, and Jeremy Mould applied infrared techniques to the Tully–Fisher relation in 1978. Measuring galaxy brightnesses in the infrared offered a twofold gain. First,

---

[6] According to the best current data, the Universe had the last laugh, and all the old papers using a value of 75 were not at all wide of the mark.

[7] Marc Aaronson was killed in a freak accident while observing at Kitt Peak in 1987. Huchra went on to play a key role in redshift surveys, and Mould is Director of the National Optical Astronomy Observatories.

the effects of dust were much weaker, so uncertainties in how well they could correct these uncertainties would be correspondingly improved. Second, since star-forming regions are blue, the galaxy's recent history of star formation could allow a small amount of material to have a disproportionate impact on its visible light, a problem minimized by looking in infrared bands where old populations of stars are brightest. Using a new wavelength regime for the problem forced them to revisit and repeat its calibration, beginning with the nearby Andromeda galaxy. In order to measure the same region of Andromeda that they looked at for more distant galaxies, they had to remove the secondary mirror of one of the 0.4-meter telescopes at Kitt Peak, inserting an 85-mm lens and thereby employing one of the most sophisticated infrared detectors of the time on a telescope that would be at home in many back yards. They found that, indeed, the infrared Tully–Fisher relation had a tighter scatter for nearby galaxies and groups than had been seen in visible light, making it a prime tool for pursuing galaxy distances. This they proceeded to exploit with 63 galaxies in the Virgo cluster and more distant groupings. The result allowed them to map this part of the Universe, for the first time, separately in distance and in redshift.

This work by Aaronson, Huchra and Mould had a major impact – mostly on our appreciation that clusters of galaxies are dynamic structures, rather than for the distance scale itself. They established what only a few people had seen previous evidence for – that clusters continue to grow as their gravitational influence has time to attract more and more distant galaxies and groups, and that we are being pulled into the Virgo cluster at a significant rate.[8] For the Virgo cluster, their results agreed rather closely with Sandage's value (namely, 65 km/s per megaparsec). But outside our local environment, they found a Hubble constant nearly as large as de Vaucouleurs favored. Had local motions really been such a dominant factor in deriving the expansion rate of the whole Universe?

8.3 Old uncertainties and new facts

In order to test for cosmic deceleration using galaxies, their evolution in brightness must be known. We are looking across such vast distances that the lookback times are comparable to the ages of many stars, so galaxies cannot help having changed in the meantime. Sandage had calculated this effect, based on estimates of which stars contributed most to the visible light. Then he gave a talk at the University of Texas in 1967, at which a young woman, one of the graduate students, announced that his cosmological conclusions were all wrong. Beatrice Tinsley burst on the astronomical scene by applying stellar evolution anew to the question of galaxy evolution, and found that the mix of stars contributing most of the light was not what most had earlier assumed. This meant that typical galaxies, in particular the giant elliptical galaxies in clusters that were favored for tracing cosmic history, must fade a great deal over the timespans at issue. This implied that the high-redshift Hubble diagram measured not

---

[8] But the work on cosmic acceleration shortly described has thrown doubt on whether we will ever in fact join the Virgo cluster.

cosmology, but galaxy evolution. So persuasive were her arguments, and so fundamental the results, that they brought about a major shift in what astronomers wanted to measure in the deep Universe. Her influence continues today among generations of astronomers who never knew her.[9] After an astonishing career at Texas and Yale, Beatrice Tinsley died at age 40, after a four-year struggle with melanoma. In selecting a paper from her dissertation work as one of the highlights for a centennial edition of the *Astrophysical Journal*, general editor Rob Kennicutt wrote "Thereafter, galaxy evolution, not cosmology, became the *raison d'etre* for deep surveys of the Universe."

Sandage himself recognized this; there is no use arguing with data. In a pair of authoritative review articles in 1988, he first set out the "classical" tests of cosmology which he had done so much to develop, then showed why uncertainties in galaxy evolution rendered their interpretation too uncertain to base a Universe view on the results. Anyone who has been such a towering figure for so long eventually makes a tempting target for young, ambitious scientists. By the early 1980s, Sandage said that he tired of young Turks trying to make a name by correcting something he did twenty years earlier.

In the 1980s, Sandage turned his long-range plans to stellar archaeology, tracing the history of the Milky Way by a detailed survey of the spectra and motions of nearby stars. This would be an effective use of the 100-inch Hooker telescope on Mt. Wilson, even in the presence of light pollution from the Los Angeles area, and a program which he anticipated would run for much of his career. Then, in 1985, the management of the Carnegie Observatories announced that they would close the telescope to direct resources to their 2.5-meter telescope in the pristine skies of Cerro Las Campanas, Chile, and toward development of yet more powerful instruments. This was, in a way, a palace coup, with younger scientists trying to move the observatory in new directions. Sandage was, understandably, personally affronted. His closest ally in the turmoil seemed to be, oddly enough, Halton Arp. After the other staff astronomers let it be known that Arp's long-term quest to demonstrate that some large redshifts were not due to cosmic expansion was unproductive, and that this project would not receive even the usual staff allotment of telescope time, Arp felt himself all but exiled. He ended up in Garching, outside of Munich, and continued work with numerous colleagues, taking advantage of the new scientific atmosphere and incoming X-ray data from the *ROSAT* program. As Dennis Overbye describes it, Sandage and Arp decided that they had remained friends, even during the years when their scientific differences had been so deep that they hadn't been on speaking terms.

Meanwhile, the traditional route to the Hubble constant was pursued by ever-widening groups of astronomers, culminating in one of the handful of Key Projects designated for large-scale studies with the Hubble Space Telescope. This project was so large that it called for three principal investigators on two conti-

---

[9] Tinsley also had no mean reputation as a teacher. I once ran into a bartender at a Pizza Hut location in The Hague who still recalled having taken an introductory astronomy course from her.

nents – Wendy Freedman of the Carnegie Observatories in Pasadena (organizational heir to Mount Wilson), Rob Kennicutt at the University of Arizona, and Jeremy Mould of the Mt. Stromlo and Siding Spring Observatories in Australia. The plan was straightforward in concept, but called for great care in execution. *Hubble* would not be able to measure Cepheid variable stars in galaxies that lay beyond the possibility that their redshifts include a substantial component from local motion, but it could range perhaps ten times as distant as ground-based instruments. Its major advantage was the razor-sharp images it delivered above the atmosphere, dramatically reducing the problem of Cepheids appearing blended with an unknown number of surrounding stars, which would bias their average brightness. In addition, its orbital vantage point offered a flexibility of scheduling no longer dictated by the phases of the Moon, so the sequence of observations could be tailored for the most complete detection of Cepheids in a quantifiable range of pulsation periods. The selection of galaxies was a major part of the proposal – even for a Key Project, with hundreds of orbits of observation allocated, each distance measurement was very expensive and precious. The final selection included galaxies which had hosted the Type Ia supernovae long considered as "standard bombs" for distance estimates; galaxies in small groups with other kinds of galaxy not hosting Cepheids; and a representative group of galaxies in the nearby Virgo cluster, which could serve to calibrate distance estimators based on global galaxy properties. The first set of post-refurbishment *Hubble* images made it plain that the telescope could meet the goals of these observations, and indeed it immediately began to yield images showing Cepheids far beyond the reach of ground-based telescope (Fig. 8.11).

The nearer galaxies would also serve as a cross-check with distances from the Tully–Fisher relation, as well as with three newly developed techniques for measuring galaxy distances, each of which showed great promise with ground-based equipment. In a field rich with such designations as IUE, KPNO, VLA, ESO, SDSS, and CDM, it will not be surprising that these techniques are often described as PNLF, GCLF, and SBF.

PNLF denotes the Planetary-Nebula Luminosity Function. Planetary nebulae (Fig. 8.8) are puffs of gas, lasting perhaps 10,000 years, that were once the outer layers of dying sunlike stars. Once the nebula has dispersed, the remaining core of the star is a white dwarf (Chapter 3). The nebulae shine as they absorb the copious ultraviolet light of the hot stellar core, and reprocess much of this light into a few spectral emission lines. The strongest of these is an oxygen line at 5007 Å. This emission is so strong that ground-based telescopes with appropriate filters can pick these objects out of the glowing background of stars in galaxies as distant as the Virgo cluster, 60 million light-years away. Straightforward calculations of how these nebulae fade with time, on a timescale which is more historical than astronomical, indicates that the statistics of their luminosity measured in the oxygen line – their *luminosity function* – should depend little on the initial chemistry of the stars or their mix of ages, and thus might provide a good way to measure distances. In fact, surveys in neighboring galaxies whose distances are well known bear this out. George Jacoby of Kitt Peak and Robin Ciardullo of Penn State University have been particularly active in establishing this technique, which, using a phase in the life of old stars, can be used on any kind of galaxy, and is in fact least subject to confusion by other kinds of nebulae in the elliptical galaxies that contain only old stellar populations. This stands in contrast to Cepheid variable stars, which are short-lived and massive, and are therefore found only in galaxies with active star formation. They are plentiful in spiral and irregular galaxies, but unknown in ellipticals.

Fig. 8.8. A nearby planetary nebula, Messier 27 in Vulpecula. These clouds, puffed off by many Sunlike stars as their cores are shrinking to become white dwarfs, have consistent statistical properties which allow them to be used for distance estimates to nearby galaxies. Image obtained with the 4-meter Mayall telescope of Kitt Peak National Observatory.

A similar statistical comparison underlies the GCLF method – the Globular-Cluster Luminosity Function. Globular clusters (Fig. 8.9) are very rich balls of stars, sometimes over a million, found clumped spherically around the cores of spiral and elliptical galaxies. Our galaxy hosts perhaps 200 of these, while giant ellipticals in the centers of clusters (such as M87 in the Virgo Cluster) may have 10,000 (Fig. 8.10). Except perhaps in some colliding galaxies that can still form them, globular clusters are uniformly comprised of old stars. A further uniformity is found in their brightness distribution. The number of clusters in a given galaxy versus their luminosity, over and over, produces the same bell-shaped curve, with, as far as we can tell, the same mean value. Unlike planetary nebulae, we do not have a compelling theoretical explanation for this uniformity, but the data certainly point to it. Distances derived from taking this mean luminosity as a standard unit correlate well with those from other techniques, and can reach father. Globular clusters are bright enough for us to see them far beyond the reach of either the Cepheid or planetary-nebula approaches.

The third of these techniques uses Surface-Brightness Fluctuations (SBFs). This technique works only on galaxies lacking bright star-forming regions, using the statistics of the very numerous but dim old stars in elliptical and the related flattened S0 galaxies. First championed by John Tonry (now at the University of Hawaii), it measures the number of stars per unit area contributing various amounts of light. In a nearby galaxy, a given amount of light per detector pixel (or other resolution element, say if the atmosphere sets a larger limit) needs only a small number of stars. Therefore, random fluctuations from one pixel to the next are relatively large. For a distant galaxy, the same amount of light per pixel encompasses many more stars, giving smaller fluctuations from one pixel to the next. This approach requires high-quality and well-calibrated data, since we look for fluctuations which could mimic poor data quality. It also relies on the statistics of stars' brightness being the same in the galaxies we compare, which is one reason it applies only to elliptical galaxies with old stellar populations of (as far as we can tell) roughly the same age. Furthermore, we must mask regions affected by not only foreground stars, but star clusters in the galaxies under study.

What has been impressive is a comparison of the results of these three new-comer techniques. Not only do they generally agree as to the distances of galaxies, they agree in considerable detail. They agree, for example, as to which galaxies are on the front side of the Virgo cluster and which ones are on the back side. However, they are still not completely free of the distance ladder, needing to rely on distances determined for a few nearby reference galaxies to confirm their scales.

The Key Project team was not alone in using *Hubble* to tackle Edwin Hubble's program with Cepheid variables. Having spent considerable time preparing an atlas of the galaxies most suitable for distance measures with the new telescope, Allan Sandage was instrumental in several additional programs scheduled to add key galaxy distances, particularly for galaxies which had hosted supernovae, as a sort of retroactive calibration. In fact, some of the same people were tapped by both teams, for their expertise in the art and science of measuring stars in crowded images. By now, it may be no surprise that the teams came up with answers that differed by much more than their stated errors. The Key Project analysis suggested a number close to the old fence-sitters' mean – as

Fig. 8.9. One of the Milky Way's 150 catalogued globular star clusters. This is Messier 22 in Sagittarius, about 10,000 light-years from the Sun toward the Galactic Center. The statistical properties of these clusters are similar in galaxies of many types, making them useful for distance measurement. Image by Lisa Frattare and the author, obtained with the 2.1-meter telescope of Kitt Peak National Observatory.

more galaxies were observed, values from 68–71 km/s per megaparsec resulted (Fig. 8.12). But Sandage's team, from a parallel analysis, derived much lower values.

The higher value fitted well with what we were learning from completely different kinds of data (such as the microwave background, as explained below). Most astronomers figured at this point that *Hubble* had fulfilled one of its major science goals: measuring the Hubble constant with an accuracy of ±10%. Convincing the entire astronomical community that this had been accomplished was another matter. It was the completely different nature of the supporting data

Fig. 8.10. Globular clusters in a distant galaxy. This image, from the Keck II telescope, shows globular clusters as starlike points swarming around an elliptical galaxy in the Hercules cluster, about 500 million light-years distant. Bright elliptical galaxies may have thousands of globular clusters, making them amenable to distance measures based on the statistics of individual clusters. (Data courtesy of John Blakeslee.)

that brought on an unprecedented consensus among cosmologists as to the scale of the Universe, and set the stage for a completely new view of its history.

8.4 Deep background

A different route to the parameters of cosmology stretched before us, starting with the discovery, forty years ago, of the cosmic microwave background radiation. This came as one of the discoveries in radio astronomy which were not only serendipitous, but came from Bell Labs (indeed, radio astronomy itself had been such a serendipitous Bell discovery, by Karl Jansky in 1933). Arno Penzias and Robert Wilson set out to use a horn antenna near the lab's location in Holmdel, New Jersey, to carry out various projects in radio astronomy. This antenna had

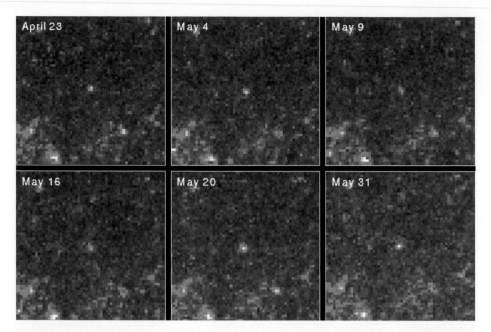

Fig. 8.11. An image sequence showing the variation of one of the Cepheids discovered in M100 (NGC 4321) over a 38-day period in 1994. The Key Project team selected this large spiral galaxy in the Virgo cluster roughly 50 million light-years away, as the first of their targets for observation. (Dr. Wendy Freedman, Observatories of the Carnegie Institution of Washington, and NASA.)

originally been used for satellite communication when the only communication satellites had been passive reflectors (such as the enormous aluminized balloon *Echo 2*). When active relay satellites such as *Telstar* were in operation, smaller and more flexible antennas were appropriate, freeing this one for other uses. They began with attempts to reduce radio interference, identifying various sources of radio "noise" that could compromise measurement of weak emission from the sky). Among these was one very local effect – delicately described as a whitish substance deposited by pigeons in the antenna, and complicated by the fact that these turned out to be homing pigeons.

Penzias and Wilson eventually found that there was an excess level of radio emission coming from everywhere they looked in the sky, and eliminated all non-astronomical origins. The potential significance of this excess emission became clear from discussions with colleagues at nearby Princeton University, who had been preparing a search for just such a signature based on cosmological calculations. This idea dated back to early consideration of what became the "Big Bang" cosmology, the notion that the Universe is expanding from an initial state that was very hot and very dense. The connection between matter and radiation would undergo a profound transition in such an expanding Universe. Early on, when the matter was still so hot that it was ionized and electrons roamed free of atoms, this gas would be opaque; radiation would be absorbed soon after

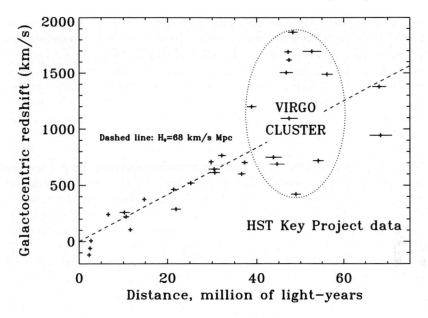

Fig. 8.12. The redshift-distance relation derived using Cepheid variable stars by the *Hubble* Key Project team. The ellipse shows the rough boundary of the nearby Virgo galaxy cluster, where internal motions add a Doppler component to the redshift.

emission. But, once the gas had cooled sufficiently, neutral atoms would form and remain, so the Universe as a whole would become transparent. Radiation emitted at this time or after would continue travelling until it encountered some discrete object which could absorb it (in our favorite case, a radio telescope). Since we look outward and see a Universe in which distance and time are mapped together, this means that there is a "wall" at great distances in which we see this transition time, and nothing beyond. This radiation, then, originated when the Universe became transparent, and it was the high density of the early Universe which should give it a blackbody character. There had been attempts to calculate the intensity (or, what is the same thing, the blackbody temperature) of this radiation – most famously a report by George Gamow whose details seem as opaque as the early Universe itself. The expected value, somewhere near 5 K, was close to the 3 K reported by the Bell scientists. This cosmological connection made this discovery, now known as the comic microwave background, very important indeed, and Penzias and Wilson were honored with a Nobel Prize in 1978.

In principle, we could learn a great deal about the early history of the Universe by examining this slice in time preserved in microwaves. Even small departures from a perfect blackbody spectrum would tell us about heating of the cosmic gas, perhaps by the first stars or more exotic objects. And if we could look closely enough, the fact that we stand here today observing it implies that it should show ripples on the sky. The formation of galaxies means that there were regions of space early on with slightly higher density than average, where gravity would eventually collect sufficient material for the galaxies we see today. This process would have started just when the Universe became transparent, the first time that matter could clump freely under gravity without being smoothed out by the radiation. With the discovery of dark matter, cosmologists realized that the dark matter could have started clumping earlier, not being pushed around by radiation, with the ordinary matter falling into its gravitational concentrations only later.

Finding galaxies at greater and greater redshifts squeezed the window of time for galaxies to form, sufficiently that it was almost guaranteed that observations with attainable precision should show these slight ripples in the microwave background. (Otherwise, as more than one astronomical wit noted, we are not here). But reports of such ripples remained mutually contradictory and inconclusive for years, because the experimental problems were formidable. Observing from the Earth's surface, the microwave background is seen behind a variable foreground of emission from the Earth's atmosphere, which must be constantly measured and subtracted. Even from stratospheric balloons, measuring the atmospheric contribution dominates experimental design. On top of this, unlike other astronomical measurements, there is no reference location we can point to without the background radiation, and absolute measurements are always more error-prone than differential ones. Thus hopes were high for definitive results from a satellite specifically built to measure the microwave background – COBE, the *Cosmic Background Explorer*. Launched in 1989, COBE used a helical scan pattern to sweep to detectors across the entire sky multiple times, with different sets of detectors measuring the spectral shape and intensity structure of the microwave background plus the Galactic dust and synchrotron radiation that could contaminate its detection.

The ripples that COBE sought come about for several reasons. Clearest, perhaps, is a timing effect. Regions with slightly more matter than average would become neutral slightly earlier than average, since the balance between radiation and matter could shift sooner as the Universe expanded. That meant that these regions became transparent slightly earlier at a higher temperature – but this is more than made up for by their higher redshift, so denser regions appear slightly cooler. How "slightly" this is became clear with the sensational release of the COBE results, showing statistically significant warmer and cooler regions at the level of a few parts per million. Initially their existence could only be shown statistically, but as the mission continued, the team had enough data to construct a reliable map of the hot and cold zone. They also found that the spectrum of the microwave background was as perfect a blackbody as they could measure, better than a laboratory source. To show the error bars on a journal illustration, they had to magnify them by 400 times.

These results only fed cosmologists' hunger for more. Detailed analysis of what we could observe from the early Universe had shown that the details of these ripples encode a wealth of information, some new and some connecting immediately to the Universe today. Most of this was in the *power spectrum* – a measure of how strong the fluctuations are measured across various angles on the sky. A large contribution to the structure of matter at that epoch came from sound. Sound – acoustic waves transmitted as regular patterns of compression and rarefaction in a gas. The universe was *different* then. This measurement would be somewhat like using a seismograph to measure the deep Earth, finding out which wavelengths propagate strongly and which ones do not. A race was on to get the first definitive data – involving satellite designers with NASA and the European Space Agency, and smaller-scale balloon payloads which could be built and launched much faster.

In fact, some of the first further results did come from a balloon experiment, BOOMERANG, which took data during a 10.5-day flight circumnavigating Antarctica, and for which the team was ready to announce results in mid-2001. But most of the publicity, and fodder for further investigation, came from the next satellite off the pad, the *Wilkinson Microwave Anisotropy Probe* or WMAP. Named after key team member and instrument scientist David Wilkinson, who died during the mission, WMAP was launched in 2001 and dispatched to a location near the L2 point, a point of gravitational stability about 3 million km outward from the Sun where spacecraft will keep the same relative positions to Sun and Earth. The results of COBE enabled some streamlining in the instrument design, notably since the mean spectrum of the background radiation was now known accurately. The improved mapping of the background radiation and its minute fluctuations (Fig. 8.13) delivered all that had been hoped for. Because various cosmological effects have different impacts on the sizes and intensities of various kinds of the acoustic signatures from the early Universe, these data alone gave values for the Hubble constant, age of the Universe, and cosmic expansion history. WMAP data favored a Hubble constant very close to the *Hubble Key Project* value (specifically, 71 km/s per megaparsec), and implied an age of the Universe of 13.7 billion years with a stated accuracy of 2%. Furthermore, its results agreed with the supernova projects (next section) that the cosmic expansion is not decelerating, but accelerating.

After the spectacular success of the COBE mission, and the further developments from WMAP, some scientists proclaimed the arrival of an era of precision cosmology, in which the field could become more like physics than the philosophy some skeptics had maintained. Supernovae of Type Ia were called into service to measure the cosmic deceleration, which was expected to give us the overall amount of dark matter and the geometry of the Universe.

8.5 The deceleration is negative

In the wake of the COBE results, the time was ripe for a new attempt to measure the cosmic deceleration that Sandage had sought for so long, now using tracers which should not change over cosmic time – or at least not as dramatically as galaxies themselves. These tracers were a particular kind of stellar explosion: Type Ia supernovae.

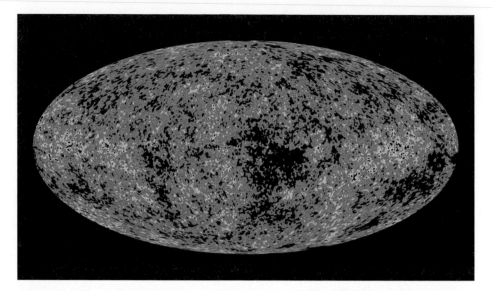

Fig. 8.13. An all-sky map of fluctuations in the cosmic microwave background, the product of the first year of measurements by the *Wilkinson Microwave Anisotropy Probe* (WMAP). Pseudocolor mapping is used to make these subtle patterns stand out more clearly. This map has been processed to remove the signatures of foreground dust and gas in the Milky Way as well as more distant radio sources. (NASA/WMAP Science Team.) This figure appears in color as Plate 8.

Supernovae come in several flavors, which we can distinguish observationally based on their spectra and history of fading after the explosion. The most numerous are Types Ia and II. We can distinguish these immediately by their spectra (Fig. 8.14). Type Ia supernovae show very broad spectral troughs from outflowing gas, with heavy elements such as silicon, sulphur, calcium and iron most prominent in the optical range. In contrast, Type II events are dominated by hydrogen lines. Fitting these facts into detailed calculations of what stars are made of and how much energy they could release has led to very different schemes for how these two kinds of explosion happen. Type II supernovae (like SN 1987A in the Large Magellanic Cloud) represent the final act in the life of a massive star, triggered by the collapse of the star when an iron core forms, creation of a neutron star (or possibly, sometimes, black hole) at the center, and subsequent demolition of the rest of the star driven by the shock wave and neutrino flash from that event. The hydrogen in the spectrum comes from the outer atmosphere of the star, which had escaped involvement in the nuclear reactions taking place deep within.

Type Ia supernovae are quite different. Such a supernova represents the destruction of a white dwarf when an accumulation of mass from a close binary companion pushes it over the Chandrasekhar mass limit. So far, simulations indicate that rather than a symmetrical collapse forming a neutron star, the results is a more irregular explosion completely destroying the white dwarf. The

progenitors may be much like ordinary novae, with the proviso that these minor explosions do not blow away all the accumulated gas, gradually increasing
the white dwarf's mass. An important point is that all the heavy elements in a
Type Ia supernova are home-grown, in the star's previous fusion or during the
explosion itself, so that it will make very little difference what the initial chemistry of the star was. This makes Type Ia supernovae very much alike – they
will represent the same amount of stellar mass (specifically, the Chandrasekhar
limit at about 1.4 solar masses) and very similar chemical composition, no matter whether the star began its life as metal-rich or metal-poor. The similarity
seems to be reflected in their energy output. It has long been known that all
Type Ia supernovae reach closely the same peak brightness, attracting attention
as potential ways to measure distances. At their peak, they outshine Type II
supernovae, so we can use them to very large distances – farther than we can
trace any other objects except entire galaxies and quasars.

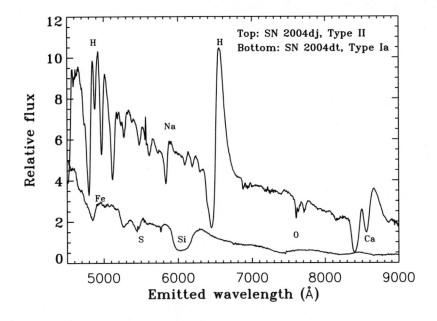

Fig. 8.14. Comparison of the spectra of supernovae of Type II, produced by the death
of a massive star, and the cosmologically useful Type Ia, produced by the detonation of
a white dwarf as gas from a companion star pushes it beyond the Chandrasekhar limit.
The Type II spectrum shows prominent hydrogen from the star's outer layers, missing
in Type Ia. Some O, Ca, and Fe lines are common to both, and labelled between them.
The strongest spectral lines show so-called P Cygni profiles, with a strong peak from
the gas all around the supernova, and a broad, blueshifted absorption trough from
outflowing material in front of it. These two supernovae were observed on the same
night in August 2004, shortly after both were seen to reach maximum light. Data taken
with the 4-meter Mayall telescope of Kitt Peak National Observatory.

In the early 1990s, two groups of astronomers set out to apply supernova distances to the problem of cosmic history. The second number of traditional cosmology represented cosmic deceleration. If we have some kind of object whose luminosity we know at the outset (ideally because we understand it thoroughly), we can ask whether its distance is greater or less than we expect from its redshift and a simple linear relation such as Hubble proposed for the nearby Universe. In a decelerating Universe, we expect distant objects to appear progressively brighter than we would otherwise expect, as we look farther away and farther back in time. The basic reason for this is that they were closer to us when the light was emitted than we would infer from the present expansion rate; it has not taken as long for the Universe to expand this much as the present rate suggests.

To the extent that their properties could be shown to be uniform enough, Type Ia supernovae might be the best way to measure this deceleration. Several groups, particularly observing from Lick Observatory and Cerro Tololo, Chile, made considerable progress on this uniformity issue in the late 1980s, systematically and repeatedly surveying regions of the sky for supernovae and following up these discoveries to measure redshifts, spectra, and peak brightness. They did find subtle differences among Type Ia events, but also made a good case that these differences were systematic enough that they could be factored into the cosmological measurements.

Two teams wound up competing to measure high-redshift supernovae and derive the curvature of spacetime. First to get organized was the Supernova Cosmology Project (SCP), headed by Saul Perlmutter at the Lawrence Berkeley Laboratory, which was exploring the best techniques for finding very distant (and faint) supernovae by 1992. Meanwhile, a more general supernova effort at Harvard began by trying to understand the systematic behavior of local Type Ia explosions, which eventually paid off in improving the accuracy of distances to faint supernovae. They organized as the High-z Supernova Search project, led by Brian Schmidt (who moved to Australia along the way). It was this group which was first to announce a stunning about-face for cosmology. The brightnesses and redshifts of their supernova sample were not consistent with *any* of the usual expectations for a Universe either so sparse in matter as to continue expanding with no deceleration, or one so full of matter as to decelerate toward collapse. Rather, their data (and soon, data from the SCP) pointed to an *accelerating* expansion.

In an urgent flurry of email as the High-z Supernova Search team decided whether their evidence was strong enough to warrant publication, supernova expert Robert Kirshner expressed how unexpected the result was. "In your heart, you know this is wrong, though your head tells you [that] you don't care and you're just reporting the observations..."

The strength of the evidence, and the strength of the reporting teams, has led to a remarkable consensus in cosmology. For perhaps the first time since relativistic cosmology was introduced, a majority in the field agree that a single set of values is a good description of our Universe, as follows. The Hubble constant is close to 70 km/s per megaparsec. The geometry of spacetime is flat, with the contributions of ordinary (4% of the critical density) and dark (27%) matter precisely offset by whatever drives the acceleration of the Universe. This gives a cosmic age of 13.7 ± 0.2 billion years.

The nature of the cosmic accelerator remains deeply mysterious, perhaps even more so than dark matter. Analogy with dark matter has led to the term "dark energy", although this remains a convenient label without any further meaning. Recent supernova observations have shown that the effects of acceleration have built up over time; early on, we do see a period of deceleration due to gravity. New information is needed to help pin down what causes the acceleration. Einstein's original cosmological constant would keep the same magnitude over time, while some suggestions driven by theory in high-energy physics suggest other behavior (either increases or decreases with time). The next aim on this branch of cosmology will to be trace the history of the acceleration in detail, to at least distinguish among these broad kinds of suggestion.

A leading contender toward this goal is a dedicated satellite designed to discover and measure large numbers of distant supernovae – so many that we can get independent measurements of the acceleration across several spans of cosmic time. This initiative, the SuperNova Acceleration Probe or (SNAP), remains in the planning stages, but has strong backing in both the physics and astrophysics communities. This would consist of a *Hubble*-class telescope equipped with a large, wide-field camera so that large areas of sky could be kept under constant observation to catch supernova on the rise. There have also been proposals to use large slices of *Hubble* time for a precursor study – proposals which are now like all *Hubble* proposals in being subject to great uncertainties in the observatory's future.

Even today, general relativity provides the basic framework in which we interpret cosmological data. Einstein's description of curved spacetime has proven robust and powerful. But is there reason to think this will continue to be so?

Further reading:

The texts of the papers from the Curtis/Shapley "debate" of 1920 may be read at http://antwrp.gsfc.nasa.gov/htmltest/gifcity/cs_nrc.html

Dennis Overbye's *Lonely Hearts of the Cosmos* (HarperPerennial, 1991) is an engaging history of cosmology on the eve of COBE and the discovery of cosmic acceleration, largely woven around the life and career of Allan Sandage.

Two sharply discordant views of the history and workings of the COBE results are given by John Mather and John Moslough, *The Very First Light: the True Inside Story of the Scientific Journey back to the Dawn of the Universe* (Basic Books, 1996); and by George Smoot and Keay Davidson, *Wrinkles in Time* (Wiliam Morrow, 1994).

An inside view of the case for an accelerating Universe is given in Robert P. Kirsher, *The Extravagant Universe: Exploding Stars, Dark Energy, and the Accelerating Cosmos*, (Princeton, 2002).

E.A. Tropp, V.Ya. Frenkel and A.D. Chernin sketch the life of a founder of relativistic cosmology in *Alexander A. Friedmann : the Man who Made the Universe Expand* (Cambridge, 1993).

9 The view from Einstein's shoulders

"If I have seen a little farther than others it is because I have stood on the shoulders of giants." This line of Newton's, as barbed as it may have been when directed to a colleague of lesser physical stature, aptly describes the way in which science, as a collective enterprise, does manage to advance. Each of us is heir to all those who have gone before. So it is with relativity. There has been a century for other great, but different, minds to explore its possibilities and consequences. So many people have contributed to its intellectual framework that it is no longer strictly Einstein's result, or his conception. In fact, it no longer matters except to historians and literary critics juist how Einstein came to his revelations or described them, since so much mathematical development and so many new applications have been worked out by others. Still, relativity and Einstein are inseparable. Could he have penetrated to ultimate truth with relativity, leaping to the final description of the Universe?

Newton's physics reigned for more than two centuries. It did more than predict the motions of worlds and outcomes of experiments to exquisite precision; it subtly set the tone for how we thought about the Universe. Metaphors of clockwork were the rage, and physicists were content to think of forces and action at a distance – but always of space and time as absolute, given properties of any point or event. As we have seen, the Universe of relativity is a bizarre inverse of Newton's, in which space and time can be rubbery and the magical distant effects of gravity are replaced by the curvature of spacetime, present and active everywhere. So far, relativity has passed every experimental and observational test to which we could subject it, with remarkable precision.

Over the last century, though, philosophers and other observers of the scientific process have drawn attention to the ways in which they see science progress, or at least change. Karl Popper drew special attention to the role of falsification in testing scientific ideas, which has generated a new orthodoxy. He also noted that, rather than an incremental process of refining older ideas, we often see changes in shifts of the entire paradigm – the whole fabric of underlying ideas. There can hardly be a more striking paradigm shift in all the history of science than the transition from Newtonian to relativistic mechanics. But history also suggests that we may never be able to arrive at the final, ultimately correct theory – only a succession of constructions which reproduce our measurements ever more closely over an ever-widening range of circumstances. So simply in a historical context, we expect relativity to someday be seen as the special limiting case of a broader understanding, just as we now see that Newtonian physics is the limiting special case of relativistic physics when all motions are much slower than light and the curvature of spacetime is small.

There are, in fact, more pressing reasons to think that relativity must be an incomplete description of physical reality. Physics in the twentieth century produced another theoretical edifice just as powerful as relativity, painstakingly erected as much more of a group effort – quantum physics. Like relativity, this theory has vast predictive power, ushers in a completely different way of viewing matter and energy, and has suggested completely unsuspected effects. But the mathematical worlds of quantum mechanics and relativity are incompatible. And there remain some subtle predictions of general relativity whose experimental verification has yet to be accomplished.

Among these are *frame dragging* and *gravitational radiation*. If gravitational attraction is a bit analogous to the attraction between different electrical charges, relativity predicts a new effect analogous to magnetism. A rotating mass will slightly drag spacetime around it, in what is called the *dragging of inertial frames*. Near a black hole, this means that an object would eventually be unable to remain stationary no matter how powerful the forces pushing it. Closer to home, a satellite in a completely undisturbed orbit around the Earth will see its reference frame slightly deflected with respect to the distant Universe as time goes by. This effect is now being tested with extraordinary precision by the *Gravity Probe-B* spacecraft (GP-B).[1] GP-B uses four extraordinarily precise quartz spheres as gyroscopes, comparing their reference frame to the external frame provided by a reference star. These spheres are electromagnetically suspended out of contact with the spacecraft structure, and an identical sphere is suspended as a reference so the satellite can correct its path for drag and similar effects. The gyroscopes are shielded from the Earth's magnetic field to eliminate this source of interference as well. General relativity predicts that the Earth's rotation will produce a difference between these two nominally inertial reference frames, amounting to 0.04 arcseconds per year on GP-B's orbit. The star IM Pegasi was selected both for sky location and for its radio emission, so that very-long-baseline radio observations could pin down its precise position and motion in the sky before using it for such a fundamental purpose. The satellite is also intended to measure a larger relativistic effect, the geodetic effect, a twist of its reference system simply because it moves in curved rather than flat spacetime. This effect has been measured for the Earth–Moon system using laser ranging of reflectors placed on the lunar surface, but GP-B is designed to measure the magnitude of this effect to one part in 10,000.

The results from GP-B may be known by the time this book sees print. This mission is not only a technological triumph, but a testament to the persistence and vision of principal investigator Francis Everitt of Stanford, who has been one of a handful of scientists shepherding this idea through theoretical development,

---

[1] Gravity Probe-A contained a maser atomic clock on a sounding rocket, which flew a 115-minute trajectory from the Virginia coast on June 18, 1976. Its data confirmed the equivalence principle, by showing that the clock's rate depended on its location in the Earth's spacetime "dimple".

technical challenges, and repeated threats of cancellation for nearly four decades. What became GP-B was initially proposed by Leonard Schiff at Stanford, and George Pugh of the Department of Defense, in 1960. NASA funding of the project dates to 1964 – surely a longevity record for any individual program.

Gravitational radiation results from the basic premise of general relativity: matter tells spacetime how to curve, and the curvature of spacetime tells matter how to move. But as the matter moves, it creates a mismatch with the space curvature, and thus ripples in spacetime moving away at the speed of light as the curvature corrects itself. Sufficiently large masses moving fast enough should be detectable across cosmic distances. Such masses might include asymmetrically collapsing stellar cores, binary neutron stars, and the mergers of black holes. Attempts to detect these waves to date have been unsuccessful, which is not surprising since the sources we know about would have been much too weak for these experiments to detect.

A gravitational wave passing through an object will produce a pattern of changes in spacetime which have the effect of squeezing it alternately in the two perpendicular directions that are themselves at right angles to the wave's direction of travel. For the sources we might consider in deep space, these distortions will be tiny. Their measurement is made even more delicate by the fact that our measuring rods, even if made of light, will be changed by that same distortion, so the changes we measure are even smaller second-order effects caused by the distortions having an effect which is not quite symmetric (analogous to the arithmetic problem about why the time-averaged speed over some path is not the same as the distance-averaged speed).

To be sure, there is strong evidence that we have seen astrophysical effects of gravitational radiation. General relativity predicts that a pair of masses will slowly lose energy to gravitational radiation, with the effect more important for greater masses and tighter separations. Some binaries of white dwarf stars are tight enough for the effect to be important, but even white dwarfs are so large that potential tidal effects, interchanging energy between the orbit and stellar interiors, could be important and poorly known. Binary systems consisting of neutron stars or black holes would provide much clearer tests, and pulsars would be especially useful because of the accuracy of timing their pulses can provide.

This accounts for the excitement following the discovery first of a pulsar (PSR 1913+16) in a binary system with a quiet neutron star, by Taylor and Hulse,[2] and more recently a binary pair of pulsars (PSR J0737-3039) by Marta Burgay and Duncan Lorimer. Such binaries should lose energy via gravitational radiation fast enough for us to measure the decrease in their orbital periods as they slowly spiral together – and measurements show that they do, at just the rate predicted for this effect. The pulsar–pulsar system shows other effects as well, such as the delay on one pulsar's signal caused by traversing the "extra" curved space near the other when they pass closest together from our point of view. However, its orbital decay has yet to be measured. This orbital decay, in the few systems in which we can see it, provides us with secondhand evidence of

---

[2] Rating a Nobel Prize in 1993 – this time taking care to include both parties.

gravitational radiation, but it would be a much more complete test if we could measure these waves as they sweep over the Earth.

Instruments with a good chance of detecting gravitational radiation, at least from the kinds of cosmic objects we are sure exist, must be enormous projects. This comes about from the small magnitude of the effects and concomitant large distances over which they must be measured. Approaching completion at two locations in the states of Washington and Louisiana are the two detectors of LIGO – the Laser Interferometric Gravitational-wave Observatory. Each detector includes two 4-kilometer vacuum cylinders situated at right angles to one another, with lasers being reflected multiple times along each cylinder. The laser beam is allowed to interfere with itself after these reflections, giving a measurement which is exquisitely sensitive to tiny changes in the path length. The two arms and twin detectors are essential to weed out other effects (including measurable disturbances from vehicles some distance away in Louisiana). LIGO is also part of an emerging global network of detectors, especially in Europe and Japan. Its sensitivity is being continually increased as the team gains experience in reducing interfering effects, so that within a few years we should see the gravitational wave patterns from colliding neutron-star and black-hole binaries within a large slice of the Universe.

Though gravitational-wave astronomy is still lacking in actual detections for analysis, plans are underway for even more sensitive systems. The LISA project (Laser Interferometry Space Antenna), under development by NASA and the European Space Agency, would entail multiple spacecraft in independent orbits, measuring their separations using laser interferometry. Periodic changes in their separations would betray gravitational waves, to much larger wavelengths than we can probe with detectors distributed all over the Earth. The isolation of space removes a host of compromising effects ever-present on Earth. Although none of these systems has achieved a detection, history suggests that most of us would put our money on Einstein.

The successes of relativity are many and undeniable. One vocal critic of relativistic cosmology, Jerry Jensen, began a conference presentation by giving relativity its due: "The successes of Einstein's 1905 papers, and the theory of General Relativity published eleven years later, cannot be overemphasized: the orbit of Mercury, the displacement of stars near the Sun during a total eclipse, the time delay in accelerated atomic clocks, and the retarded atomic particle–decomposition rates in high energy physics. These predictions have all been verified. We have looked for and found neutron stars, event horizons, gravitational lenses, and we have exploded nuclear bombs. Has any scientific theory ever been more successful?"

So where should we look to go beyond relativity? Any yet more general theory must predict the things we already see, not only as understood by relativity but quantum mechanics. Einstein spent the final decades of his life seeking such a "unified field theory", today known as a Grand Unified Theory or Theory of Everything (GUT or TOE). Attempts so far to see how gravity might be quantized will work in two dimensions, but not three or four. This is an area where something completely new – such as string theory which replaces all the entities of previous theories with vibrating one-dimensional objects – is likely to be the road forward.

It is not only scientists who seek to go beyond Einstein. Since its initial publication, something about relativity has attracted naysayers, many of them amateur scientists (and some outright deserving of the term "crank"). What is it about relativity that attracts denial of theories which are supported every time someone uses a GPS receiver?

Unfortunately, it has not been only physicists who took from Einstein the model of how physics looks. Pure cranks also set out to completely rewrite the conceptual basis of physics regularly, introducing new mathematical formalities and inventing new words wholesale. However, they do not begin with Einstein's grasp of what we do, in fact, securely know. Lots of us in academia get books, papers, diskettes, CDs, especially after getting our names on articles that come to public attention. [3] Many people clearly care a lot about the abstruse issues of philosophical import involved in relativity. Its conclusion do run contrary to every expectation of everyday experience, and the mathematical framework ranges from abstruse to incomprehensible. It is no wonder, perhaps, that so many people still think that this cannot be the way the Universe works.

With the worldwide spread not only of education, but of data and ideas flowing with the speed of electrons, the resources needed to investigate our Universe are available to more people than ever before. In his *Outline of History*, H.G. Wells noted that history shows us clusters of brilliant minds, grouped in time and space, and concludes that this is tragedy – reminding us of the wasted potential of all those who did not have the fortune to be in an environment where they could flourish. And W.E.B. DuBois wrote that "ability and genius are strangely catholic in their tastes, regard no color line or racial inheritance. They occur here, there, everywhere, without rule or reason." At least in the limited spheres of the flow of information and access to ideas, our increasingly global society is becoming more egalitarian with every new Internet connection.

In this connected world, where should we look for the insights to go beyond Einstein's view? From a graduate student whose fascination with the philosophy of physics overshadows any interest in the minutiae of electromagnetism, to the professor's dismay? From a woman in the Navaho nation, blending ancestral world views with the other sources available in this digital age? From an amateur scientist fascinated with the possibility that slight tracking anomalies in the *Pioneer 10* and *11* spacecraft may be the first sign of genuinely new physical processes? From a computer scientist in Bangalore, who suddenly sees deep analogies between mathematical abstractions and hitherto unconnected physical laws? From another physicist working in Switzerland, this time poring over masses of data from high-energy particle collisions and muttering, "That's funny..."?

We are surrounded today by an astronomically vast heap of sterile ideas and misinformation in this most basic of sciences. But someday, there will be a gem

---

[3] Please do not add to my collection on this topic. I can provide testimonials from my old professors on my complete mathematical inability to review them properly.

shining in this morass, from a worthy heir to Newton and Einstein. Wherever the next revolution comes from, I certainly hope to be able to recognize it, wherever and whoever its origin. It can scarcely help setting the stage for a further century of breathtaking discovery.

Further reading:

The September 2004 issue of *Scientific American* included a series of articles on tests of the limits of relativity and the kinds of theories that might supersede it. Brian Greene's *The Fabric of the Cosmos: Space, Time, and the Texture of Reality* (Alfred A. Knopf, 2004) gives a recent explanation of the quest to make string theory the next revolution in our view of the physical world.

Marcia Bartusiak traces the quest to detect gravitational waves in *Einstein's Unfinished Symphony: Listening to the Sounds of Space-Time* (Berkley Publishing, 2003).

Index